Jungle Combat

A Combat Pilot's Tape Recorded Transcripts from Vietnam
1968-1969

Colonel John "Ace" Astle, U.S. Marine Corps (Ret.)
Combat Pilot, Vietnam War and Desert Storm Veteran
Presidential Helicopter Pilot, Medical Evacuation Pilot
31 Air Medals, 2 Purple Hearts and Presidential Service Badge
Former State Senator of Annapolis, Maryland's 30th District

Gemma M. Jablonski

For more information please go to www.AuthorGemma.com
and/or email: Press@AuthorGemma.com, Info@AuthorGemma.com or Agent@AuthorGemma.com

Cover design: Ilovemycover.com
Formatting: Polgarus Studio

Printed in the United States of America First printing, 2021

Dedication

This is focused on those of us who have served in Vietnam. Many of us were born during WWII and others in the late 40s. We all had family members who served in the military during that war. I think we all can remember the parades honoring those men and women who had won a war on two fronts. It was a war that the whole country was committed to. So as we began to come of age it was assumed in many communities that we would serve. I know that was how it was in the community that I grew up in. It was considered the honorable thing to do.

We had a vision of what war was like. We had heard the stories from our older family members. Of course these were told in such a way that their activities seemed glorious. So that was the mindset that many of us had as we got off the plane and stepped onto the soil of Vietnam. We were quick to discover that war is not glorious. It is a very ugly thing filled with fear, death, destruction and deep emotions as we lost friends. One of the first things to go is patriotism and mom and apple pie. What replaces it is an intense loyalty to ones friends. We had a little saying...better dead then look bad. Letting down ones' friends was just not done and often people would do very courageous things to protect friends.

While we were there engaged in what for many of us was a life and death struggle, there were changes going on back home. Young men burning their draft card and mass demonstrations where they chanted: "…hell no, we won't go." Young men leaving to go live in Canada. It was difficult for us to understand because it wasn't the country that we had left. I remember hearing that students were picketing a Dow Chemical recruiter on a campus because Dow made Napalm. We knew nape was a terrible weapon but it worked to save many American lives. We also had to operate under rules of engagement.

That seemed a little crazy. Here we are engaging an enemy who doesn't operate with those rules and we felt it put us at a disadvantage.

Having endured and survived, we came back to a country that blamed the war on us. People called us baby killers and worse. In some places we were shunned. I got home feeling that I had accomplished something with my life. I had served my fellow Marines and saved lives. As I was walking in uniform along the concourse of the LA airport I was spit on by a woman. I wanted very much to smack her in the face but that would not have looked good on the part of a Marine officer. So for a long time we dealt with our memories, with our scars both physical and emotional. We did alone or in small groups of friends who had shared the same experience. But for the country as a whole we were forgotten. That is what makes a recognition ceremony so special because all of us are being recognized for what we accomplished.

I want to close with this. While we were there fighting and many young men were doing their best to avoid the military there was some push back on the part of many of the vets.

There is a saying that is engraved on brass plaques that I'm sure are hanging in many homes today; I know it hangs in my home. It goes like this—"War is an ugly thing, but not the ugliest of things: The decayed and degraded state of moral and patriotic feeling which thinks that nothing is worth war is much worse. A man who has nothing for which he is willing to fight; nothing he cares about more than his own personal safety; is a miserable creature who has no chance of being free, unless made and kept so by the exertions of better men than himself."

— John Stuart Mill

My brothers, we are those better men. Welcome Home!

Colonel John "Ace" Astle, U.S. Marine Corps (Ret.)

Contents

Prologue

"Many autobiographical books have been written about the Vietnam experience, most of them years after the fact based upon imperfect memory. By contrast, the contents of this book reflect immediate reactions to the grim reality I encountered in Vietnam at the time on a daily basis.

It's all here, from the drama of facing danger in combat, to the frustration of dealing with the military bureaucracy, to the excesses of exhausted Marines in drunken revelry. What follows is a chronological transcript of those tapes edited only for coherence, and not to protect the readers' sensibilities. It reflects my extreme swings of mood —from fear and anxiety to grief and exhaustion to relief and exhilaration. The events described are sometimes barbarous, sometimes ridiculous and sometimes sublime, but they are always real. In the theater of war, life gushes forward at a heightened pace. But amid the cacophony and confusion, the discerning reader will hear the earnest sounds of a young man emerging into manhood. Be forewarned the language is sometimes raw and cruel, for which I make no apology. It is the honest language of who we were then, an audible snapshot of that hardest of times."

- Colonel John "Ace" Astle U.S. Marines (Ret.)

Forward

“The Vietnam war became my reality when I stepped off the USS Buckner, a troop ship left over from WWII, and on to a landing craft in QHON Harbor located in central Vietnam in the summer of 1965. My unit was part of The First Cavalry (Airmobile). I was a young Captain fully of P & V, a term used by the Army meaning combat ready and anxious to get into combat.

I had received my commission from Morgan State College, after Infantry Officer’s Basic Course, I deployed to Germany as part of the First Infantry Division stationed in Berlin. That, followed by a tour with the Third US Infantry (The Old Guard), where I served as Commander of the Tomb of the Unknown platoon.

The First Cavalry Division was created and designed specifically for its ability to rapidly deploy combat force in a Jungle environment using helicopters. This concept was soon tested in combat during the first major battle involving conventional forces early in November of 1965. This battle took place in the Ia Drang Valley located in the central highland. North Vietnamese launched an attack on units of The First Cavalry located at LZ X-Ray and Albany. The use of helicopters in roles of resupply, fire support, evacuating wounded and command/control was key to the success of troops on the ground.

My admiration for the heroes who flew those helicopters was taken to another level after reading the accounts of the day to day life of a Combat pilot.

The war changed my life as it did so many of my fellow warriors and like John, went on to make a major contribution to our way of seeing the world. Our strong sense of what is important and what is BS, allows us to be able to grow old gracefully.

A dear friend, whose funeral I will be attending tomorrow,
Joe Galloway, often said:
"We may not be the greatest generation, but by God,
we're the greatest of our GENERATION!"

Lieutenant Colonel George Forrest U.S. Army (Ret.)
Battle of Ia Drang, Commanding Officer
of Alpha Company, 1st Battalion, 5th Cavalry
"We Were Soldiers Once...and Young"
- Book by Co-Author Joseph L. Galloway and later made into a movie.

Introduction

In the spring of 1968, as a newly designated Marine Aviator, I embarked for Vietnam where I would spend a year flying helicopter support missions to combat troops in the field. It was to prove a definitive period in my life, when an extraordinary range of sensation and experience was jammed into a short time span.

I was raised in Barboursville, West Virginia, a small town of some 2,000 souls perched on hilly ground near the junction of the Mud and Guyandotte Rivers, about half way between the state capital of Charleston and the city of Huntington which abuts the Ohio River. Barboursville was and remains a community of proud mountain people which has committed its sons to the nation's conflicts since the earliest days of our republic.

In frontier times, Barboursville was a bustling center of commerce where farmers from miles around congregated to trade their produce for manufactured goods. There were mills, blacksmith shops, hardware dealers, and churches of many sects, and there was politics. Barboursville was for a time the seat of Cabell County and Old Hickory himself, Andrew Jackson, addressed the citizens there during his first campaign for the White House.

Economic development in the latter 19th century bypassed Barboursville, leaving it a remote backwater town of modest pretensions, but the pride of tradition remained firmly planted. As a boy, I recall many rapt hours sitting around the hardware store and the barbershop listening to the older men speak of World War II. Beside the main road into town, proud local citizens erected a large and solemn Honor Roll of those who had served, with special asterisks to honor those who gave their lives. There were no memorials to the Korean conflict, but there were veterans aplenty with stories of their own.

Thus, I was bred amid a palpable sense of patriotic duty to military service.

As children, we played a variety of army games. I recall it was always a generic army, with scant distinction afforded different branches of the service. Within our limited context, Air Force, Navy or Marines were all Army, and it was simply assumed we would all one day take our turns to serve our country in one of them. It was less discussed than taken for granted, an unspoken presumption. So it was that when the Vietnam conflict beckoned, I answered the call. The decision was born not of any quest for glory, or special sense of mission. Rather, I went simply because everyone including my family and myself, assumed I would.

Upon embarking for Vietnam, I considered I would have difficulty making myself put pen to paper to keep the folks at home apprised of my situation, and that much unnecessary anxiety might ensue. Therefore, I determined to keep an audio diary on a little tape recorder I purchased for that purpose, and to send the tapes home to my family. My parents took readily to this scheme, and in turn sent tapes to me, tapes that served both to provide me with fond memories of home, and also as an endless source of merriment for my Squadron mates.

My mother, as mothers will do, kept the tapes and eventually returned them to me many years later. I let them sit in a closet for a long time, as I tried to put the war behind me and get on with my life. It wasn't until the autumn of 1990 that I was able to come to terms with the ghosts of that intense drama, and to listen to my voice from so many years before. It was an extraordinary experience I shared with a few close friends. Prompted by their urging, I had the tapes transcribed.

Chapter 1
The Quiet Before the Storm

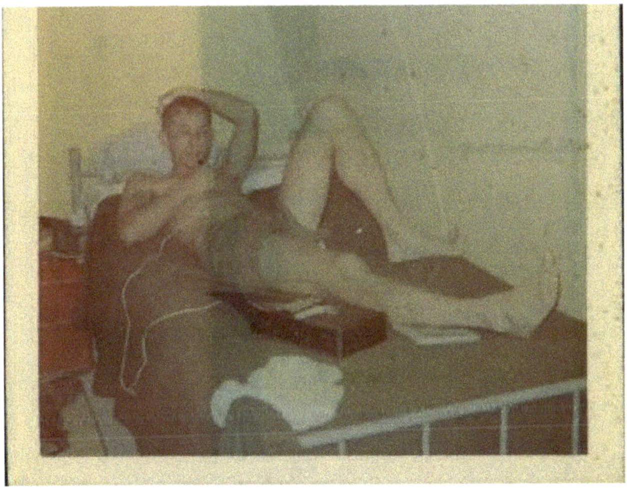

Well folks, it's me, your son John sitting here in the fabulous fun spot of the South Pacific, Okinawa. Just thought I'd take this opportunity to give you a run-down on the progress of my trip so far. Saturday afternoon, Dave T. and his wife took me over to Norton and deposited me there at the Transient Facility. I had a very unpleasant six-hour wait until the flight left. I flew on Continental with about 164 other people, plus some nice stewardesses, to Honolulu, where we had about a 20 minute stopover. Then we reboarded and had about a nine-hour and 15 minute leg, getting into this God-forsaken place about five in the morning on Saturday. This was actually something new to me because all of a sudden I discovered I had missed Friday in there someplace.

I latched up with an old staff NCO who had been recently commissioned and was now carrying the rank of First Lieutenant. He had something like 19 years in the Corps and had been in Okinawa before. He knew his way around the Marine Corps system pretty well, so I stayed with him through the initial check-in. They told us there were some other things we should be doing, but we decided they could just hang it in their ear. We went in, changed into our civvies, and headed out to see Okinawa.

We boarded a civilian bus which was a real experience since these people do not speak English. They speak Japanese or at least some dialect of Japanese. We traveled from one end of the island to the other and got off in a city called Kōsa. I think that was its name, anyway. I'm not even sure. We got off the bus and did some shopping. My pal pointed out some of the spots of interest he had visited on some of his earlier stop-offs through here on his way to Vietnam. Then we went over to the Futenma Marine Air Facility, did a bit of shopping and got some lunch in the officers' club.

They've got a pretty nice club up there sitting on a hill overlooking a bay. The Pacific is truly a beautiful piece of water. It's got a fine, distinct color to it. Also, the weather that day was pretty nice. There were scattered clouds that sort of painted a real pretty scene. We sat there, did a couple of drinks and just soaked it all in. After a while, we left and went back to town. By the way, while I was doing all that shopping, I picked up five watches in one of the little shops. Got a pretty good deal; stuff over here is unbelievable. We got them for about $20 a piece. - A special number one deal for Nisan! - Nisan is a word I picked up. It's Japanese for boy. Nasan is for girl. And, dozo is for please. And arigato means thank you. A few words like that may come in handy. Anyway, we went to the town called Kōsa, started at one end and walked all the way through it. I guess it was about a mile and a half long. We spot-checked some of the establishments and had a couple of cold beers. It was hot outside, real hot. When we got tired of that, we got back on another bus and started back to Camp Hansen where we are currently billeted. That was an interesting trip because school kids were just out of school and packed on the bus. They looked us up and down like we were from another planet. You would think they had seen plenty round eyes by now, but maybe not. Anyway, we looked at them funny right back. There's a little town just outside the base gate that sort of sprouted up there to serve the base, so we're told. We did a quick tour of it and came back into the base. Knocked around, had dinner. I ran into several old friends from Pensacola who are on their way over. We chased around the club for a while. Then another guy and I headed out into town to sample some night life here on the island. Well, that one trip was enough for me. It was, well, let's just say I'm not real impressed with these people. For one thing, it bothers me that I can't speak their language. You never know what they're talking about, and I can sense some hostility. Maybe that's just my imagination, but I don't think so. I can't be sure. Anyway, we got back about 3 a.m. which made for about 24 hours I had been without sleep. So I was pretty happy to get some rack time.

This morning we got up and went over to check in again. We have to sign in twice a day. Went up to the club, got some lunch, sat around reading magazines. Then we went up to the camp theater about one o'clock and took in a John Wayne flick. It helped our morale a little bit, I guess. By the way, before we came here, we stopped by and got the gamma globulin shot, the one that is supposed to protect you from anything that comes through a virus, like infectious hepatitis, three-day measles, and stuff like that. The threat is real enough. They shoot 500 cc's of that stuff in the cheek of your ass. It feels as though someone put a golf ball in your back pocket. I'm sure it'll go away soon enough, though.

You'll probably be interested in the attitude of the people around this place. It's amazing. I've never run across anything quite like it before. Most of the enlisted people who have just come here from Vietnam immediately go into town looking for trouble. Their care factor is so low it's really unbelievable. And the people just getting here, the enlisted people, are just herded about like cattle. It was difficult gauging their attitude because I didn't spend much time around them. They don't get off the base much, and are generally confined here.

But I have talked with many of their officers who are getting ready to go over, and watched them in the club, and I've never in my life seen anything to match it. The care factor is without doubt the lowest I've ever encountered among any group of people. They sit up there in the club and you never saw money flow so freely, or the booze either, for that matter. It's just amazing to sit there and watch those people, a difficult sensation to describe. Most of these guys do not want to go to war, and are really upset about it. I don't know whether it's the fear of going to war that's bothering them, or whether the fact that they're leaving their parents, or maybe they just don't give a shit. I don't know; maybe it's a little bit of all of that that makes their attitude the way it is. It's interesting, though. I guess we'll go up to the club again tonight. They do have a movie up there which kind of helps give you a chance to sit there and get your mind off your problems.

I'm scheduled to pull out of here Wednesday at five o'clock. I don't know what time that will be back home where you are. I guess it'll be the 16th instead of the 17th here, because I think we're one day ahead of you. It will be Wednesday, the 17th of June, 1968, at 5:30, I am to board a plane at Kadena Air Force Base bound for the Republic of South Vietnam. I'll fly into Da Nang and from there I'll get my assignment. I don't have any idea yet exactly where I'll be going. I'll be sent to a group someplace and then the group will break down the assignment and put me in a Squadron.

Not to change the subject but one of the really good things that has happened since I've been here is the fact that I've gotten to see a lot of my old friends. People I hadn't seen for a long time. It kind of helps you getting together with your friends when you're all pretty much in the same boat. It really helps a lot just to sit around and bullshit about it. Makes you feel a bit better, anyway. Well, I guess I better sign off on this thing now.

The guys are up at the club waiting for me. We're going to catch a cocktail before dinner, get something to eat and then I suppose hang around and watch the movie. Then I guess we'll grab some sleep. I'll toss you another line before I get in the country.

You know, it's kind of funny. I've never tried to write a letter quite this way. And, it's really weird sitting here in the head talking; seems like I'm talking to myself. So please don't try to read anything into the tone of my voice. It's just that this is kind of a new and strange experience for me. I mean, the experience of just sitting here talking into this thing. As far as the way I feel, my attitude right now, I'm a little bit apprehensive, but at the same time I'm a little bit excited. I'm not quite as down in the dumps as a lot of my friends are. I guess I have the facility of looking for something good to come out of everything that I've ever experienced. But all in all I think my attitude's pretty good. I've been staying out of trouble. I have blown quite a bit of money, but other than that, things are okay. So I'll talk to you when I get in the country and maybe one of these days I'll get the hang of writing my letters this way.

June 17, 1968

Well, here I am, another day, another dollar. Another glorious day to be alive in the Corps. Things are getting pretty boring. I had an attitude check this morning and decided that I was just bored to hell. It really gets tiresome just sitting around here with nothing to do. Went down this morning to check in and I was about five minutes late. So I had to go see some First Lieutenant and he gave me a big ration of shit. I just grinned back at him and said to myself - yessir, yessir, three bags full. Because it means so much that I sign this list, you see. It's kind of a slap in the face anyway. They don't even trust us to stick about the base here, so we have to sign this muster list twice a day.

But I guess that's life in the Corps. Went in last night to sign this muster sheet. We were about a half hour early, so the damn people made us stand around for 30 minutes just to sign this muster sheet. The word had been put out it couldn't be signed until 1600 hours. Course that really torqued us.

Well, today after I went down there and talked with those wonderful people, I came back up to my room and managed to get both of those packages wrapped. Those are the watches I'm going to send home, and I guess this afternoon I'll run 'em down to the post office and get 'em in the mail. Then I think I'll take a magazine up to the club to read because it's air conditioned up there, a lot nicer than down here in the BOQ. The BOQ is nothing more than a barracks with two-man rooms, and the humidity of this place is simply out of the question. The minute you step out of air conditioning, it's like stepping into a big pot of glue. The air is so sticky and hot. Today is even worse than usual because it's clouded up and looks like rain. That brings the humidity up even higher.

Well, another one of my friends shipped out of here this morning. He left from Kadena AFB about five o'clock. So I guess I won't be going until

Wednesday the 19th. I'm supposed to leave here about 5:30 in the evening. In the meantime I will be bored stiff sitting around here doing nothing. The movies help a little bit, but they're only shown twice a day on weekends, and once a day in the work week here. Everything's on a seven-day schedule. They're really pushing all the time. In fact, I've never seen anything quite like this since I've been in the Marine Corps. One would think I would be really happy at the opportunity to get out for a sight-see around the island, but as I said the other day, I could really care less about what's going on outside the gate here. I'm not really impressed with the place. I can see how if a man had to stay here 13 months he'd just go out of his ever-loving mind. Well, I guess I really can't think of anything else to say today, so I'll go ahead and close out. Maybe I'll find something to tell you later on this afternoon after I eat lunch and everything. But I think I'll spend most of the day up there in the air conditioned sanctuary of the club. By the way, last night I went up to the club and slept about an hour in the sitting room up there. I was pretty tired, too tired to even go into the bar. Just went in to watch a movie, which occupied my mind a while. Came back here about 10 and slept in to about nine this morning. I didn't really want to sleep that much, but at least when I'm sleeping, I'm not bored. Well, so much for now.

June 18, 1968

Hi folks, this is John again. Here I am sitting in the comfortable rack they gave me in the BOQ. Had a pretty good afternoon as I said, I went up the club with a magazine. Read it and got in line for a haircut. Getting a haircut here is quite an experience. They have women barbers. They don't speak English very well, so there's little of the usual barbershop chit-chat. But once they finish cutting your hair, they give you one of the damndest rubdowns you ever saw. I thought that lady was going to twist my head off. Massage your face, neck and shoulders. It's really something.

After that, I knocked around, saw a movie, mailed those packages, and wandered over to the transient facility to sign the muster sheet. A funny thing happened as I was walking down the street. I saw this bus go by and I heard a guy yell out the window – “Hey, John Astle!” I looked around and couldn’t make it out, but I noticed the bus was stopping so I went to see who it was who was yelling at me. These two guys come walking up, and all of a sudden I recognized a face. It was Sergeant W. of the Huntington Recruiting Station. He’s over here and is leaving on Wednesday, same time I am. He’s assigned to the First Marine Air Wing. So I’ll probably run into him again. So then I went on over to the mail since President Johnson lowered the duty free amount. It’s down to $10. That’s kind of a pisser, I think. Then we went on over to sign the damned list. Talked to the baby sitters there a few minutes and wandered on back to the club. Had a pretty good dinner – Salisbury steak, French fries and gravy. The food’s not all that bad here, and they have these little Okinawan women running around tossing it out. Some of them are kind of funny. They don’t speak much English. It’s a lot of fun to kid around with them. Then I’ve been sitting around here talking to another guy. He went on up to the club, and I just thought I’d get a few more words off to you. Then I’m going back up there. They’re gonna have a floor show tonight. It’s gonna be some guy. I was hoping for a woman, but no such luck. I think I’m getting the hang of writing these letters this way. In fact, I’m kind of enjoying talking into this thing. Now I just can’t seem to stay away from it, and I’m sure I’m gonna enjoy tossing these things off.

My roommate went into town tonight. He wanted to get a steam bath and a rubdown and anything else he could find. He tried to get me to go along, but I tell you, after that first night I chased around in town, I’m pretty much fed up with these locals. I don’t even have any desire to walk out of the camp gate. I’d just as soon stay around with the round eyes who speak my language.

By the way, tonight when I went down to sign the muster sheet and check on my flight, we discovered that my flight’s gonna get in sometime during the night. We’re supposed to leave here about 7:30, so they say, and it’s about a five-hour flight. So my roommate, the one with 19 years in the Corps, he and I decided that if we don’t land in the middle of a mortar or a rocket attack, we’re gonna be really disappointed. He said if that didn’t happen, he’d put in for a transfer right away to someplace where the action is happening. I don’t know about that personally, that’s what he said. But it should be interesting to see what kind of greeting we get when we come rolling into that wonderful place called Da Nang.

By the way, you can tell Jim I said Hi, and the same to anyone else you run into. Tell them I'm over here basking in the warm sunlight of the western Pacific in that fabulous fun spot called Okinawa.

You know, I think I've finally discovered just why I don't send letters to too many people. As you've probably noticed, I just get these little bits of conversation down in snatches. I think my problem is that when I sit down, I can't think of a whole lot to say. Maybe if I were to dictate it to somebody, it might fill up a page. Then as I sat there thinking about it, and trying to decide how to say it better if I throw some stuff out as not very important, and then I get down to only two or three sentences left and figure, hell, there's not enough here to worry with. So I never seem to get anything done. But I think with this tape recorder, I'll be able to sit down when something comes into mind worth telling you about, and then shut it down until something else comes along. Cause with this thing you don't feel as though you have to sit down in one sitting and whip the whole thing off at once like you do with a letter. So this should work out pretty well for all of us.

Oh, by the way, I got my moustache started. I think I'm going to grow myself a handlebar. So when it gets long enough to start twisting I may write home and see if I can't get you all to scrounge me up some moustache wax so I can wax that beast down and get her twirled at the ends. It ought to be kind of nice to bring home.

Well, I guess I'll let this be all I've got to say for today. I'll probably think of something to say tomorrow. So in the meantime, I think I'll buzz back up to the air conditioned sanctuary of the club, take in that floor show. Waste a few more hours. You know, it's kind of funny, but before I left home, it seemed like time was valuable. Now all of a sudden it doesn't seem to mean a thing. In fact, if it weren't for this watch that tells what day it is, I wouldn't even know. Hours and days all seem to run together into one big blob. Well, I'm off to the club. I doubt if anything interesting will happen tonight so I'll check back in with you tomorrow.

June 19, 1968

Well, hi folks. This is John again, and it's Wednesday, June 19. It's been raining here the last couple of days. The weather hasn't been too good, but my roommate and I decided that today we're going to take a trip over to Vietnam just to see what's going on over there, see if we can't help get the situation straightened out. We're tired of those people over there messing around trying to keep the war going just so they can draw their combat pay. Besides, we'd like to get some of that combat pay too. Well, I've got most of my packing done now, and in a few minutes I'm going to carry my gear on up to the club, and then go down and check out of the BOQ. We leave from the club, so I guess I'll spend most of the day until about two or three o'clock this afternoon hanging around up there. I'll probably end up in the game room working the slot machines. I've been up there two days in a row, put in about $25 and taken out about $30 so that's not too bad, I guess. That much fun and $5 ahead. But I've been working that one-armed bandit so hard I've got blisters on my hand. Listen, if you get a chance when you send your first CARE package over, I wonder if you could include a small pair of scissors that I can use to trim my moustache when it gets to the stage that it needs trimming. A razor's a little bit bulky to do it with, but I think a pair of scissors would work well enough.

I ran into two more guys last night that had been in my class at Pensacola. They just came in yesterday and won't be leaving until next Monday, so they have about five days to wait around this place. We sat around and watched the movie last night, and then played some ping-pong, a little shuffleboard and worked the slots a few minutes. Then we sat around and talked some about war and stuff.

Well, I'm gonna close out for now and pack this thing up and get all my gear hauled up there to the Officers' Club, so the next time I have anything to say to you it will be from good old RVN.

Chapter 2

Welcome to Vietnam

Well, folks, here I am in good RVN. Today is Thursday, June 20, and I've had good fortune. I am stationed at Marble Mountain, which is part of the Da Nang complex, and it looks as if life's going to be pretty good here. They have a mess hall and serve ice cream with all our meals. They have a real nice Officers' Club, all in the same stone building and I'm living right on the beach.

I checked into the Squadron today and I'm assigned to HMM-164. It's the 64th Squadron and one of my buddies from Pensacola is stationed in the same Squadron. In fact, he and I are living together now in the same hootch. These hootches are just kind of thrown together, shacks with screen wire all the way around so that they can be left open. It's really sort of like an open room with some metal beds in it. Ours is pretty sparse for shelving or closets, or anything like that, but I think with enough time I will be able to accumulate enough gear to make life fairly comfortable.

I got into Da Nang last night about ten o'clock and spent the night at the R&R center. The first thing I noticed when I got here was the humidity level which is just incredible. I thought Okinawa was bad but that was nothing compared to this. And the mercury got up to 104 or so and everybody was saying the hot season hasn't even started yet. I have never encountered a place this hot and humid; it is completely out of the question.

Everybody says things have been going pretty good around here lately. They haven't been catching that much shit, and things have been pretty easy. But I don't know. The word's out that the Viet Cong are going to launch a mortar attack or a rocket attack on Da Nang Village, which is just a couple of miles away and they're afraid some rounds might fall into Marble Mountain, or some of the other complexes in this area. Of course, we do have bunkers right outside the hootches here. It's just a matter of a hop, skip and a good sized leap and you're in a bunker. Also, they're built pretty strongly, which is good to know.

I checked in with disbursing today to find out how my money would be handled. I'll get the checks here when we get paid, and then I'll put them in the mail to you. I'll make them payable to you, or whatever. Then you can put them in the bank for me and we'll proceed from there. Now I may keep out a small amount for myself, enough to subside on or subsist on I should say and that way you won't have to send me any checks at all. You can just go ahead and handle my financial responsibilities for me.

I also arranged to have income taxes taken out from what's left over from the $500 so that at the end of this taxable year, I won't have to pay that much income tax. It'll be pretty much taken care of already, so I won't have to send the IRS a hellacious check at the end of the year.

Everybody I've talked to says life seems to go pretty fast over here that the days go pretty quick. The only thing I've noticed is that it seems like a year ago that I left home.

One thing I have noticed here is the distinct sound of artillery fire in the distance. In fact, last night as we flew into Da Nang, we looked out the windows and could see the flashes of bursting rounds in the distance. But when we got into Da Nang, everything was pretty quiet. There didn't seem to be any problem at all.

By the way, I think another thing I would like you to send me are some sheets. It seems that sheets are in short supply around here and in fact, the rack I'm sleeping on now doesn't have any. I'm sleeping on a bare mattress. It isn't a very clean mattress either, though I get the impression cleanliness isn't a real big thing around here. Most everybody is wearing smelly flight suits and if I keep on sweating like I am now, my whole body will be pretty smelly before long.

One thing I will say about Vietnam is that I don't think I'm going to like it very much, and will probably be happy to end my tour and get back to the States. Most of the people who've been here a while don't seem to sweat so much and maybe I won't either once I get acclimated. I have started taking salt pills again and of course that makes you sweat a whole lot more. Also, I've noticed that my appetite's fallen off. I guess we get so intent on drinking fluids that we're not really hungry when dinner time rolls around.

Right now I can hear artillery fire in the distance. I guess that's about the only thing around here to remind me there's a war on. Otherwise, this place is like being back in the States, except for the fact that every place you look there are sandbag bunkers and barbed wire and everybody's carrying loaded guns around.

Well, I can't think of too much more to say right now, so I'll just go ahead and close out. I gotta wander across the street and make a head call. So I imagine tomorrow I'll have more to add. And once I get flying, I should be full of new adventures and the experiences that I'm having over here so I'll talk to you again tomorrow.

June 21, 1968

Well, folks, here's your son John again, sitting in the head in that lovely, fabulous fun spot of Southeast Asia, Marble Mountain. Actually, this is one of your more interesting heads. It's an outdoor type, sits right alongside the street. You can almost reach out and touch cars going by. Well, more jeeps and trucks than cars. The insides are screened so that you can watch the world even as the world is watching you.

Interestingly enough, I took my first shower yesterday which proved to be rather interesting because as I was standing there soaping myself down, in walked a woman. One of the old Vietnamese mamasans. Well, I smiled very politely and said Hello. She said hi and went about her business.

Last night I went to the club for a while and then I ran into some old friends of mine and they put me onto a movie in one of the hootches. So we left the club and went on over there. They had 14 pornographic movies. Really an unexpected treat. Of course, being one of the hard core, I stayed until the bitter end.

Oh, by the way, when you send my first CARE package, I wonder if you would include a marking pencil of some sort. None of my skivvies are marked with my name, and if I have these mamasans that work here do my laundry, I'll never get my stuff back again. So I definitely need to put my name on my stuff. But I can't get anything so obvious at the PX here, which is just useless. They don't have anything that you need. We have to have most everything we need sent from home, thank you very much.

Well, I started flying today. Quite an experience. My first hop was just to get familiar with everything. We just went up, flew around a bounce pattern a little bit, went over the bay a little bit, orbited around a while and came back in. Sort of an administrative hop. I came back in and was supposed to go back out on a test hop, but all of a sudden they told me I was going to go out on a fly-by. Well, the first hop I didn't get to fly too much so I thought maybe on the second one I might be able to take the stick for a while. So I took off and we went out over the water and orbited for pretty close to 45 minutes. They had two Hueys, two H-34s, two H-46s, and two 53s. You know, the H-53s. And then besides that they had two F-4s, two A-6s, two F-8s, and two A-4s. Some kind of General was having something going on and they wanted all these aircraft flying over to impress him. So I did get about 10 or 20 minutes of stick time while I was orbiting around the circle.

Then when I got back from that, they up and asked me if I'd like to go back out again. I said sure, I'd fly. Then I discovered the hop was out into bad guy land. So I strapped on my pistol and they gave me this piece of gear that must weigh close to 30 or 40 pounds. It's nothing but a big piece of armor they call a bullet-bouncer. It covers your chest and everything, and it will stop a 50 caliber round. And then you get in there in those armor-plated seats and you feel pretty secure.

Well, we went out into one zone and picked up some people and went over to another zone and picked up a couple other people and took them over to another zone. Out there we could look around and see jets making strafing and bombing runs on places out in the distance, and we flew close to some artillery pieces that were sending out fire. It was kind of interesting. It was enjoyable flying because there wasn't any real danger attached to it, but some of the guys in the Squadron went out today and managed to take some fire. In fact, one Major, the S-3 officer in our Squadron who's a real dipshit, anyway, he and two other 46s were out going into a zone and they started taking a little bit of fire. Well, they'd called for some Hueys to come in. The Hueys are the gunships that give us fire suppression while we're going in and out of these zones. Well, they hadn't come up yet so the Major got on the radio and called up his two other birds and told them he was going down to draw their fire.

Well, that is what he did all right. The whole world opened up on him. Somehow, though, he managed to skate by without taking any rounds. But we were wondering how he managed to skate by. It was pretty hairy.

I'd planned on going out to the beach today. As I told you earlier, the beach lies just across the street. But they decided that flying was more important than beach time. So I missed out on it but I think I'll get some beach time in soon. The Squadron is going to have a party tomorrow afternoon so we won't be flying then and I should be able to get some beach time.

I don't think I'll be flying in the morning. I flew most of the afternoon today so I don't think they'll be scheduling me in the morning. So I should be able to get some sun in the morning, which sounds like a pretty good deal.

I bought some film for my camera the other day and the only thing I could get was slide film. They're pretty short of film over here, among other things, so I think what I'm gonna do is just take out the roll and send it on home for you to have developed. And by the way, those other pictures that I left with you, when you get those developed, I'd appreciate it if you'd send them to me so I can see them.

Well, I'll talk to you in a minute here. (Pause) I saw somebody coming down the street there and I felt kind of silly, you know, sitting in the head talking to myself. Or at least that's the way it must seem to someone walking by. Well, I was just looking at this tape thing and it looks like I have a good bit of talking left to do.

It looks like rain now so I'm gonna take a break from this and get on back to my hootch. I'll catch you tomorrow. Adios until then.

June 23, 1968

Well, today's Sunday and the only way I knew it is Sunday was by looking at my watch. I flew my first combat mission today but I was kind of disappointed – I didn't get shot at. In fact, I didn't see anything at all like there was a war going on over here. I went out this morning with one of the Majors in the Squadron. We took a two-bird flight out to a place called An Hoa. I don't know if you've ever heard of it. I don't think I had though the name has a familiar ring. Anyway, we were supposed to go up there and pick up a Colonel and take him somewhere else. But we waited around for about two hours until some General came in and the Colonel ended up leaving with the General in his helicopter. So we lifted out and we got a call after we lifted out that they needed some troops lifted out of one place over to another place called Thong Duc.

Well, first off, the Major said he wouldn't do it unless he got some Huey gun support. When they said we couldn't have any, he told them to forget it, he wasn't going in at all. Then we discovered that we had two other birds in our Squadron head over there, so the Major said that would provide enough firepower to protect ourselves, so we went after all. But of course nothing happened. A routine run in, we dropped off our people and lifted out. All in all, I got in about three hours of flight time today and five hours of admin time yesterday.

Another one of my old buddies came into the Squadron today, a guy I'd known in the training command and at basic school. That makes four of us altogether in the Squadron now, the new kids as we were called. After the morning crew finished flying today, the other Squadron took over the afternoon runs, we had a Squadron party on the beach. As is the custom, all the officers from the CO on down got thrown in the beer bin along with all the beer cans and the ice by the enlisted men. It's some kind of custom, apparently, so I went along with it. Then we had some food and some beer and then we walked down to the beach and played around in the water and then had a big football game. Since I am one of the new people, they threw me into the water an extra time. It was really kind of a pleasant afternoon. It kind of makes you wonder, you know. It was so peaceful and beautiful along the water you can forget there's a war going on.

Well, I'm going to close off for now and get some rack time. I'll be talking to you tomorrow.

June 24, 1968

One thing I noticed about Monday mornings over here is you don't normally have the Monday morning blues because you don't have a weekend to get all worked up for it. Monday morning was just like Sunday morning which was just like Saturday morning, and so forth and so on.

We didn't fly this morning, but I had to go in and meet with the group CO, some Colonel. He gave us a lecture. I had to go with all the other new guys to learn what was expected of us over here. Then he welcomed us and said how wonderful it was to have us in his command. Later, back at the Squadron, I found out I was to be the afternoon test co-pilot.

So after lunch I went back to the Squadron, picked up a magazine on the way and sat around there the rest of the afternoon. Got about two hours of rack time. As it happened, I sat there from about 12:15 to about 5:00 and never did go flying.

Well, yesterday evening, after I spoke to you, I decided to move out of the hootch I was in and in with two other guys I knew well in Pensacola. Then this evening another guy moved in so now there's four of us. Three of us were together at basic school, and the other guy was with us while we were in Pensacola, so we've got a pretty nice little group. We're starting to fix up the hootch now. Went out today and got some light bulbs and got some stuff from the carpenter's shed.

So I got all that fixed up and this evening I unpacked my suitcase and got everything hung up and everything I need put in the locker. It's a pretty nice setup. I've got a light that hangs down from the ceiling right over my head. Of course, it's a non-shaded light. I'm going to try to make a shade out of an old tin can. There's a little shelf behind my head to put stuff on. So with the exception of not having any sheets, it's a bit more comfortable.

Well, it seems I'm always talking to you from the head. But I just finished my business and am going on back to the hootch. I thought maybe if the guys aren't too busy, I'll just sit in there and talk to you. It's kind of funny sitting in front of people talking into this thing. But these are pretty good guys, and I know them pretty well so I don't feel so bad about it.

If you listen carefully you'll hear the sounds of war. (Background music from a radio) Sounds like war is really hell, doesn't it? Well, right now I'm looking at one of my roommates, Tom W., sitting on his rack wearing his Mickey Mouse ears. One thing about pilots in the Marine Corps is they have a great sense of humor.

Well, I guess tomorrow's going to be another action-packed day with thrilling adventures. I'm gonna fly a test hop tomorrow if I'm really lucky. My moustache is coming along pretty well. I'm gonna try to save it and wear it home so you can all see what a real combat veteran looks like.

There's one other thing I wanted you to include in that first CARE package. I went off and forgot my alarm clock and I think that I might need one over here. Sometimes we have to get up pretty early, and without an alarm clock, that can be difficult to do. When I first moved into this hootch, we had quite a battle because it was already occupied with about three million roaches, this according to actual body count, but due to perseverance and superior weapons systems, we were able to overcome the roaches. Lying about on the floor now, as I say by actual body count, are three million roaches.

This evening we made up a roster on the wall for medals. You can get a medal for killing 20 roaches. You can also get a similar roach medal for killing 5 roaches with your bare feet. So far, Tom W.'s in the lead with two, Ray D. and I are tied for second with one a piece and good old Tom B. is bringing up the rear with a big goose egg.

Our hootch is pretty centrally located about three blocks from the shower, a block and a half from the shitter, 150 yards to the pisser, and only about two blocks to the mess and the club. I'm really taking this war bit seriously. In fact, I've got my pistol hanging over my bed within close reach. It's fully loaded with five rounds in the chamber for instant action. Having loaded guns handy is a great morale booster.

It's kind of funny, but as I look around my hootch now I can see all my roommates writing letters in the old fashioned way with paper and pen. But here I sit writing letters with the new method, which is a lot easier and less trouble. Of course, you all know my affinity for the new ways, especially those that make less trouble. Well, I think I'm going to go and take a shower and wash the dirt and grime from a hard day's combat from my body so I can smell sweet when I climb into my rack without the sheet. And by the way, I don't have a pillow either nor a blanket. Like I said before, war's hell and sometimes we have to make sacrifices. But I do have three roommates to keep me warm and dry and cozy. And we are all up tight, right Tom?

"Right John."

Right Ray?

"Right John."

Right Tom?

"Huh? Oh, right, right."

(Sounds of gunfire.)

What you have just heard is the sound of perimeter guards firing tonight. Ordinary stuff, really. Today was pretty normal. This morning I read a skin magazine. After lunch, I went down to the Squadron because I had been designated test co-pilot for the afternoon. Well, I was sitting there thumbing through another skin book waiting for something to happen when the Operations Duty Officer came over and said he'd like me to do a turn-up. Well, a turn-up is normally done by one pilot. That's all that's needed to go out and turn the aircraft up. And all you're doing is turning the aircraft up so the maintenance people can make various checks on the systems while the aircraft is actually operating, and the engines and rotors and everything going.

Well, as he was talking to me, one of the heavies walked in and the Duty Officer decided to check it out with the heavy first since I wasn't really fully qualified to do the job. Well, the heavy decided that I shouldn't do it. Then a regular test pilot walked in, the one I was supposed to fly with, so he took me out and ran me through a turn-up. When we finished that, we went out and took another airplane, did a test hop on it, came back, and they had yet another airplane that needed a turn-up. So I went out and I ran it up myself.

Now, this in itself might not seem like a great accomplishment, but to those of us in the know, it's quite a thing. It represents the first time anyone has given you any real responsibility and you can't really believe it has finally happened. You tell yourself it can't be for real. Here they are actually letting you go out all by yourself and turn-up the airplane, even start the rotors. Well, it gives you quite a feeling after you've gotten it turned up and you've shut it down, and everything's still in one piece and working like it's supposed to. Also, as the turn-up pilot today, I had an opportunity to make my first decision as a pilot. When I was starting to turn-up the aircraft, I got the number one engine started and noticed the temperature indicator, what we refer to as the T-5, wasn't operating. So I immediately shut down the engine and called the crew chief and told him I wouldn't turn-up the aircraft until we had the T-5 working properly. So he called an avionics man out, we got the T-5 indicator fixed, and then they were going to balance the Stability Augmentation System, the SAS. After that he was going to have me do a topping check, which is a check we do to measure the power output of each individual engine. Well, that SAS was so screwed up that we had to shut it down, and they finally decided they had to take the whole SAS unit out and replace it.

So, that was about all that really happened today. After I finished my last turn-up, I grabbed some chow, and now I'm sitting in the head talking to you.

I'm beginning to get settled into the Squadron now. It's taken a week or so, I guess, but I'm beginning to be able to remember the names of the guys in the Squadron. I've had the opportunity to fly with several of them, and I can say I am flying with a really great bunch of guys. I've said it before and I say it again, I've never found a better bunch of guys than the people I've been fortunate enough to work with in the Marine Corps.

Another interesting thing happened today; another one of our friends from the training command reported into our Squadron and will be moving into our hootch, although I notice they have not been able to get him a regular bed. It looks like he will be sleeping on a folding cot, which is kind of bad. The rest of us have metal beds with mattresses.

Well, you know I've only been gone from home about 14 days, but it seems like years ago I left home, and already I can see that when the time comes to return home, it will be the greatest experience of my life. I really think I will be psyched up for it after 13 months in this place. I may get settled in, all right, but I'm afraid that after a while...

Well, this is just an unnatural environment we're living in here. Of course, I guess you all realize that, too. But I'm sure I will be fully happy to get home. In fact, I've already noticed that I really miss looking around and seeing women. There's a lot of Vietnamese women working around here but they don't really do much for me. Well, I can't think of too much else to say so I guess I'll close off for now. It looks like I'll be able to get this tape finished tomorrow and get it into the mail to you. And then maybe I'll start getting some mail from you all.

Chapter 3
Baptism by Fire

Today is Wednesday, June 26, 1968. After I finished that first tape last night, nothing else much happened so I just went to bed. I guess it must have been about 1 a.m. but early this morning, around 4:30, they came in and woke me up, said I had a 5:30 brief. I was pretty sleepy but got up, sort of half-way shaved and went over for some breakfast.

By the way, I've started to eat breakfast every day now, a strange custom I didn't have at home. Oh well, maybe I'll gain some weight while I'm over here, which would be all right. Anyway, we got a brief from the intelligence officer and while we were there, we asked what the problem was last night. He said about 25 Viet Cong had walked into an ambush that the friendlies had set up out there. They think the VC were out to set up mortars to hit Marble Mountain, but we got most of them. At least I hope they got most of them. I guess we'll be finding out soon enough. Anyhow, they sent us out taking about 80,000 pounds of gear and ammo to some of our units scattered around in a place called Happy Valley. Well, the first trip was perhaps the hairiest external load I have ever carried. By external I mean we had it hooked to the outside of the helicopter by means of a hook and pendant, or what we call a pendant. We carry it to the zone, put it down, push a button in the cockpit – this is called pickling the load, to release it on the ground, and then we lift on out.

Anyway, the first load took two planes. We were the first two to go into Happy Valley today, so we were a bit apprehensive, though we did have two Hueys on call to give us support if we needed it. Well, it seems as though we had to drop this load of ammunition into a small outpost that was nestled on the side of a mountain, a zone about 1,300 feet above sea level. The General canopy of the jungle was about 75 feet off the ground – I mean the trees and everything. So we had to hook the load by means of a 100-foot pendant.

Now, this is not really all that good because in order to even pick the load up off the deck, we had to hover at just over 100 feet. This is not real good because we don't have single engine capability over here because of the adverse conditions – the heat and moisture and all. So if you lose an engine over here, you're just out of luck. Well, anyway we picked up the load, the two planes did. We were flying wing and the other guy was flying lead. We went down into the zone, the Hueys made a couple of sweeps to see if they could draw any fire. They couldn't so the friendlies popped some smoke from their zone. Now this zone, the hole they had cut in the trees, was only about 20 feet in diameter. It's a pretty difficult job to lower a load in a helicopter on the end of a 100-foot pendant when the pilot cannot even see this hole. He merely goes on the commands of his crew chief who's looking out through the bottom and barking instructions – go forward, left, right, up and down, to guide him into the zone.

Well, the first bird finally got his load in there, and then we went in right behind him. But once we got our load down, instead of just pickling it, we had to hover there and pick up some captured gear they had to take back for intelligence evaluation. Needless to say, we were all a little bit tense while all this was going on, but it worked out pretty well. No special problems. We came on back and then went back out flying regular externals into a small company perimeter – flying in ammunition and stuff. They had some artillery pieces set up. And for the first time since I've been over here I was treated like a real officer and I actually got to do half of the flying.

One thing interesting, though. These guys are pretty good. They've gotta be, you know, to last this long. And they never make any attempt to grab the controls from you or anything else when it looks like you might screw up something really bad. And I did. I kind of scared myself a little bit, and probably scared him too, but my HAC didn't say anything. I was coming into this zone. It was Hill 55, a central location, and I was making an approach from the north. Now it is just a little hill, maybe 55 meters above sea level, but the zone is on the south side of the crest and we had to approach from the north because the wind was from the south.

In order to get into the zone, we had to overfly the crest of the hill and then overfly some telephone wires that have been stretched up there, and then drop down into the zone. Well, I was coming back into the zone carrying an external, the cargo nets that they use to affix these bundles to the pendant, and I got just a little bit too low. All of a sudden I heard the crew chief yell, "look out, you're gonna hit the wire." And so, without thinking, I did a rapid jerk on the collective and pulled it up. I drooped my turns from about 100 percent to about 90.

Now when I speak of drooping turns, what I'm referring to is the bleeding off of rotor rpm, and anytime in the future when I refer to drooping, this is what I mean. Drooping is very bad, by the way, because that's the lifeblood of a helicopter. Without your turns, you can't stay in the air and somewhere between 88 and 92 percent rotor rpm, the CH-46 loses its generators which in turn causes a loss of the stability augmentation system, which really makes your job a lot harder.

Well, after that things went pretty well. We flew all morning up until about 12:30, then came back in and I filled out an after-action report. Oh, I was going to mention, we sat down on Hill 55 for lunch, went up to the mess. I went up to one of my platoon mates from Basic School, I hadn't seen the guy in a long time, and talked to him. He said he had extended and so was going home for 30 days, and then coming back over. I suppose he must like it here.

I think one of these days I'm going to take the tape recorder up in the airplane with me and try to rig the microphone outside my helmet so I can get a recording of the talking that goes on while we're in the air. It's just unbelievable. You're monitoring two radios, your FM, what we refer to as a fox mic, and also your UHF radio. There is always a bunch of conversations going on at the same time, a lot of confusion. In fact, you're also monitoring the ICS at the same time. When you're doing these externals going into the zones, whoever's flying it would just have to turn off his radios so he won't be distracted. You can't have all that garbage in your ears when your crew chief is trying to give you instructions. But I have discovered that after a while you get to the point you can decipher what's intended for you and what isn't.

Of course, a lot of the information is beneficial to you. Like sometimes somebody will call in and they'll request from one of the controlling agencies controlling the artillery fire out there if they have fire going into a particular area. If they do, they give you coordinates that the fire is impacting so you know to avoid it. It's important, because sometimes our own planes get shot down by friendly fire. So you copy those things carefully. In fact, I filled up two sheets on my kneeboard today because you have to keep track of where you go, how many people you carry, how much cargo, and so forth. You have to keep track of all the frequencies and the code names of the controlling agencies that you're working with out there. Then you're copying down these coordinates and things.

Oh, by the way, we got a MedEvac. They called in an emergency MedEvac while we were in the area so we diverted and picked it up. We took two planes. We hovered while our leader made the pickup. But that was another thing – we had to write down the coordinates where this guy was so we could go in and pick him up. So I did quite a bit of work today, though all in all it was pretty enjoyable. Well, I'll be talking to you again tomorrow.

June 27, 1968

Last night I celebrated seven days in the country, my first week and they brought it off with a bang. Actually, several big bangs. It was like this. About one in the morning we were all sleeping very soundly in our racks when suddenly we heard a giant WHOMP. Well, the first one was enough to wake us up. The second one hit and we were all wide awake. By that time we were headed toward the door and a siren went off, we had dove into the bunker. I tried to find my pants, but I couldn't. Couldn't find my boots either. It was a rocket attack. Sort of a mild one, really, and they didn't even hit our side of the compound, though two army personnel were killed and about seven wounded.

The funny thing was what we call bunker casualties. About 37 guys reported injuries incurred running to the bunkers, or coming back in the dark after the attack. Anyway, I tried to record it but in the excitement I didn't get the recorder turned on. So I kind of missed the spontaneity of on-the-spot coverage. So you'll just have to settle for this rather tamed-down version. It wasn't all that bad, really, but a little frightening. Actually, it was more frightening this morning thinking about what could have happened. At the time, we were all so sleepy and it was all so new, we weren't all that concerned. We were just more or less in a daze. I'd like to say we won't have to go through any more like that, but of course we will.

We had a briefing today. They told us new guys that that mortar attack was a good one to break in on, that we would see more and bigger ones before we were done here. But it answered one of my questions. I found out what it feels like to be under fire.

Well, today was a rather interesting day. Went in this morning at 8 a.m. for our regular all-pilots meeting and then after that we had a brief by the Executive Officer for all the new guys. He just sort of gave us a little lecture on how to be good guys and fit nicely into the scheme of things in the Squadron. We left there and had another lecture by one of the Senior Pilots on some of the techniques that are used around here to get in and out of zones and so forth.

We started out this afternoon flying bird #13, but fortunately none of us is superstitious so we just went ahead and started it up. The bird wasn't functioning all that well, but we lifted out and went out to the area we were supposed to be working. We were supposed to be part of a four-plane division, but we were so long delayed in getting our bird that the other two planes were already working by the time we got there. So we pretty much operated as a two-plane section. Got out to the zone and we asked them if they had any internals because neither bird had operating external hooks. Unfortunately, they only had one load that was internal, so we told them we'd haul that. We took the load out, came back to Marble Mountain, got a new bird and went back out again. We work from a central area, picking up loads and taking them to designated units. We did that most of the day and it was pretty routine.

We did have one interesting moment later on. We had taken a VIP out to check on his people, and were waiting on him. Meanwhile, another bird in our section had gone into Hill 55, which we were working out of, shut down and gone to get some chow. Well, we had come back in and we were going to shut down and pick them up. We were sitting in the zone and all of a sudden about a thousand meters to our direct front, which was at that time to the east, about seven or eight artillery rounds exploded. We immediately got on the radio and called the artillery controlling agent for the zone and asked him if they were firing, and he said no. In the meantime, so we heard later, the mess hall just erupted into pandemonium, everybody running around in all directions. So we immediately lifted out of the zone and went up to about 5,000 feet which is well clear of artillery.

After that we went to another zone, picked up the Company Commander who wanted to go out for a bit of aerial reconnaissance. So we flew him around a while. Then as we were coming back into his zone, we all of a sudden saw a flash of artillery exploding. But much to our comfort we discovered it was friendly fire coming out of the zone we were going into. So we called them up and they ceased fire until we got in. We ended up doing a MedEvac out of there, by which time it was almost dark. And maybe I got half an hour of night flying in.

We had some trouble with our radios. My mic wasn't working. I could hear but nobody could hear me, which wasn't all that great a problem. It was just a bit of inconvenience when I was trying to talk to the HAC. Anyway, it was eight by the time we finally shut down. So it was a pretty long day and I think before long, my buns will get callouses from sitting in one position for so long. Plus you factor in what we call the bullet bouncer which is a piece of three-quarter inch steel sewn into a little jacket-like affair which protects my torso. It'll stop a 50 caliber round, which is comforting to know. That plus your armored seat makes you feel pretty good. But all that weight tends to push you down into the seat which makes it very uncomfortable over long periods of time. Well, that's about it for today.

June 28, 1968

Today my hooch maid arrived, my hooch maid being what would amount to the cleaning lady. She comes in every day, does the wash, sweeps and all. But today I was lying on my rack reading and all of a sudden I felt the presence of somebody and I looked around and there squatted down and looking over my shoulder at this book, was a Vietnamese girl. She doesn't even speak enough English that I could say she speaks English at all. One of my friends has a Vietnamese phrase book and a few minutes later she comes over with it and points out a couple of things that had the English alongside Vietnamese characters. She wanted to learn to speak a couple of phrases. So I'd pronounce it for her and then she'd pronounce the Vietnamese for me. So I picked up a couple of words of Vietnamese.

Well, I'm supposed to be the night turn-up pilot; I just hope they don't have any aircraft they want turned up. A turn-up pilot is one that goes out on maintenance, well, I guess I explained that to you already. That's about it for today; catch you tomorrow.

I'm back. What you hear is the sound of mortars firing into the center of Marble Mountain. I'm sitting in the bunker with the rest of my roommates listening to this mortar attack. Up above you can hear the sounds of the Huey gunships that had gone up to try to search out where the mortars are coming from. It's about a quarter after two in the morning and as you can tell, I am not asleep.

June 29, 1968

We had change in command ceremonies today, and then we had a party. To me, it was pretty skosh because I didn't do anything other than work on our bunker. I just spent several hours filling up sand bags and building up our bunker so that the next time we have an attack we will be immune to mortars and the only thing that could really get us is a direct hit by a rocket. Yes, yes. If they put a rocket right into our bunker, my number is up.

Well, today while I was at the club, we had a live floor show. Actually, the show was of a nature that had it occurred in the States, none of us would have walked half a block to see it. But over here it was beautiful. It was great. It was fantastic and I loved it. We cheered them, we clapped for them, we whistled, we wanted more. It was beautiful. We really enjoyed it. They were really great. Well, as you can tell, I'm a bit under the weather so to speak.

June 30, 1968

Well, here I am in the head again. Today was Sunday, a pretty quiet day. I was the duty PIMP this afternoon. PIMP stands for Post Inspection Maintenancc Pilot. I was the turn-up pilot. This morning I just slept in like I usually do on Sundays. Of course, the only way I could tell it was Sunday was it was a red letter day on my Seiko watch.

This afternoon I went down and turned up three or four airplanes. Did one test hop. That was about it. I did get my first letter today. Also, we did some bunker improvement work. We decided our bunker is probably one of the most important pieces of real estate we have over here. So we went out and shored it up a little bit and made it a bit stronger.

Right now, Tom W. and Tom B. are trying to clean up inside the hooch. They suddenly got this bee in their bonnets that they don't want to live in a hovel anymore, that we're roughing it a little too much and they'd like to live a little bit better. So I guess I'd better sign off and sort of help them out a little bit. I'll talk to you tomorrow.

July 2, 1968

Here's John again. I didn't get to talk to you yesterday mainly because I had the duty which lasted from 8 a.m. yesterday morning until 8:00 this morning. I had to do turn-ups half the night, and I got about two hours sleep and then I had to get out and fly today. I didn't get to take a shower and I was dirty and it was pretty bad. But today I had a very interesting hop. It happened like this. We were out flying re-supply in a little place about, oh eight or ten miles out there, not very far. We'd just finished and a Huey went down out of one of the zones about fifteen miles from here, a place called An Hoa. Well, when the Huey went down, they put some Marines out there to guard it because they wanted to recover it. Well, they asked us to take the maintenance team out there and drop them in the zone. We started out and we asked for a report from the Huey people that were covering the operation, and they said they'd been taking sporadic sniper fire but that they had Marines out there to give us some cover and the Huey gunships were going to cover us while we went in and put these people down.

Well, we got in there and put the people down and we thought we were finished, so we came home. But we'd no sooner touched down than they told us that they wanted us to go back in and stand by in case they needed us to go in and pull the Marines back out.

They had two Huey gunships putting out rockets and machine gun fire to help cover, and then they had a CH 53 that was going to come in and pick up the Huey. Well, in the meantime, they started getting mortar fire in the zone and of course we were monitoring the conversations on the radio. We heard someone tell the 53 driver that they were getting mortars in the zone. He came back on and said, "Man, I know that. I'm getting out of here just as fast as I can." Which we thought was kind of humorous.

That's the sound of firing off in the distance. I don't know if you can hear it. Well, anyway we thought we were going to have to go back in and pull the Marines out. But the Hueys managed to do it. They were covering one another with the guns and two F-4s were dropping napalm and snake-eye. That's sort of a half rocket, half bomb sort of deal, pretty devastating. I wish I'd had a camera because these explosions are really beautiful, the colors and everything.

Now, tonight I thought I'd just point out to you that as I sit here on my back porch watching, oh, there went a star cluster over there to the northwest about a mile, anyway, I'm sitting out here on the back porch sort of watching the war. I don't know if you all have ever heard of the spookies? People mounted with Gatling guns that fire something like 6,000 rounds per minute. You've probably heard about them.

Anyway, off in the distance, I guess maybe seven or eight miles, they have two Spookies working tonight. They worked out in the same area last night. They were there most of the night. They just fly around in circles with fire, and this is truly something to see. You can just barely make out the rotators on the aircraft but when it starts to fire, there's a solid line of tracers going to the ground. And when you stop to figure there are five rounds between each tracer round, you have to figure there's a hell of a lot of fire power at work.

Well, things have been building up quite a bit around here the last couple of days. We see an increase in activity around the area. There was a lot of firing last night. Tonight over to the west, just a few minutes ago, we could hear some fire fights. In fact, we could see tracer rounds shooting all over the place. Some of them are even coming over our heads here. Of course, these things don't bother us much, we're used to them. They've called out the mobile reaction platoons already. People seem to think they're gonna try to penetrate here tonight. We have heard that sometime between the first and the 15th, the z*ps are supposed to hit Da Nang city. If they do that, no doubt some stray rounds will come over here. But we'll be safe in our bunkers.

Well, I think I'll close off now and get some sleep because I'm gonna have to fly tomorrow and I'm pretty tired right now. I've only had three hours sleep in the last 36, so I'm beat.

July 3, 1968

This has not been a very good day. Actually it's a day that did not start very well, and it hasn't ended very well. I said goodbye to my roommate this morning, Ray D. He went out to fly some recon, search and extracts and he didn't come back. He and 15 other people.

Of course, I realized this was going to happen sooner or later. The situation being what it is, people are going to get hurt. But I just didn't expect it to happen this soon.

Well, I debated whether to tell you all this. But I decided you read casualty reports same as anyone else and you know what is happening. I guess we all knew that when we came over here. But so far I have been really lucky. I've gone into zones where people have taken fire not five minutes after I left and yet I've never taken fire. I guess I have close to 100 sorties already, I don't know exactly, but a good many. I think I've already got one air medal. I'm starting on my second one and things are going pretty well for me.

I went out today about 1:00 to do some re-supply and didn't climb out of the cockpit until after eight tonight. So, needless to say, I'm pretty tired. We did quite a few re-supplies and then got some emergency stuff. One of our units ran into some pretty heavy contact today and we had to lift out about 140 reinforcements. Just before we went into the zone, they had Hueys working and some F-4 making runs on the area, plus some artillery called in. The troops in there were sending out mortars and artillery. And as I say, we went in there and didn't see or hear a round.

The zone we had to go into was pretty bad. I've never seen one like it. It was up on the top peak, maybe 1,000 feet straight up from the valley floor. We could make our approach coming from the east but to get out we had to lift up in about a 75-foot hover to clear the trees and there wasn't much clearance for the blade tips.

Well, we flew around that for a while and then it started getting dark and our Fixed Wing had to leave. We were getting kind of low on fuel, but they needed one more load taken out there, a load of medical supplies the troops needed bad, so we said we'd go ahead and take it for them. So, we ran it out and by that time we were really getting low on fuel. I was beginning to worry just a little bit. We made it back into Hill 55 with just under 100 pounds of fuel in each tank. You figure about six pounds per gallon, which isn't much when you have two jet engines burning.

But everything worked out. It's kind of heartwarming, you go into a zone, and the troopers come running up to ask you to take letters out. We even took some ice cream into one place today. And then the last mission we flew today, the first time I'd ever flown something like this, we brought some bodies out. Took them to a mortuary and dropped them off. That kind of bothers me a little bit. Interesting, they had them in bags, but there was a kind of funny odor in the plane. Wasn't nauseating or anything, just strange. But I guess tomorrow will be another dollar or two, so I guess I'll grab a shower and get some sleep. I'll talk to you tomorrow.

July 4, 1968

Well, I guess everybody's celebrating back home today. We ushered in the 4th of July last night with a bang. Had a rocket attack. It happened right after I finished the little note last night to you. I'd just shut off my tape recorder and was looking at a magazine before I went to sleep. I was pretty tired when I heard two rockets hit. Well, I grabbed my flashlight and tape recorder and headed for the bunker.

But at the same time they were having a floor show, one of those USO shows over at the club which was packed. Well, they didn't hear the first two rockets go off because they were making so much noise. But then some others bracketed the club and got their attention good enough. It was the first time I actually heard the rockets coming in. Actually, I think they were aiming at the club. I think they were trying to wipe out 90 percent of our officers and pilots because most of them were in there last night. I think we should just kick all these Vietnamese civilians off the base here. I'm sure that's how the VC know when all the officers will be in one place to make such a target like that. They did kill two men in the enlisted section – a rocket landed right in a hooch. Boy, I'll tell you, those things just scare the hell out of me, to be perfectly honest with you. Everybody seems to think we may have another attack tonight. I don't know, but I'll be sleeping pretty light, you can count on that.

Well, this morning I flew with the CO. I had to get up at 4 a.m. to make my brief. That was really nice, since I'd been up half the night in the bunker because of the rocket attack. But the hop went pretty well. We were lifting some Korean marines into a zone. Went real fast, took one and a half hours. Then I came back and started on my collateral duties. I'm the Assistant Coffee Mess Officer and the Assistant Publications and Directives Officer. But today, we were making our monthly net worth statement and I had to write up a letter and everything for the CO so he'll know how the mess is doing.

Then I wasn't feeling too well so I came back and hit the rack. You can bet I'll grab some sleep when I can. We understand that Jeanne Dixon, the lady who does all the predictions, said Da Nang would be overrun on the fourth with about 80 percent casualties. Well, they're gonna have to hurry. They've only got about four more hours before it's the fifth of July.

That's about it for today. Catch you tomorrow.

July 5, 1968

Well, the rocket attack or mortar attack didn't materialize last night. But this morning at 4 a.m. they came in and woke me up to go flying recon inserts and extracts. Well, that didn't do too much for my morale and on top of that, I wasn't feeling so well.

But, I went out and flew anyway, ended up getting 7.7 hours strapped in the cockpit. You got all that gear on plus the armored steel thing you wear on your chest and there you are, strapped in that seat. I'm telling you, it really wears you down. I think I've never been as tired as I have, coming back from these long hops.

Well, the S-2 thinks we're gonna get hit with a rocket attack tonight. They've been dispersing trucks around and all and I'm going to bed early. There's a show at the O Club tonight, live performers, but I'm just too tired to even make that. I think I'm going to catch a little bit of sleep in case I have to spend another night in the bunker. I'll talk to you tomorrow.

July 11, 1968

I don't remember exactly the last time I talked to you. Whenever it was, I screwed up. We thought we were going to get hit that night. The sirens blew so we went to the bunker and I set up the tape recorder to get it all. And then

I forgot about it and left it on all night and most the next day to the point the batteries died. And then when I went to get some more, I discovered there were none that size to be had in all Da Nang. But today the PX finally had some size C batteries in so here I am in business again.

Not much has happened since I last talked to you. I haven't really flown for about a week. Seems as though I got ahead on hours for the month. I had more than 21 so they're holding me back and letting some of the other co-pilots get more time in to even us all out. So I've been having things like the duty turn-up for the afternoon and also the night turn-up, which I'm stuck with again tonight, incidentally.

I'd like to say a few words about that. It's called Bald Eagle, a standby situation. We have four crews and four airplanes standing by for a night launch for MedEvacs, or anything of an emergency nature. Anything after dark is considered an emergency if you have to launch.

Well, the first two crews have to be on 15-minute alert which means they have to sleep in the ready room. Conditions there are primitive to say the least. Actually, they're non-existent. You don't get a lot of sleep. Every time the phone rings, you jump, your heart races, your adrenalin starts pumping. You know if you get called to launch, it's going to be a genuine shit sandwich and those are about the worst you can ever launch on.

I did do one thing the other day that was kind of interesting. We had an airplane go down in one of the zones with a generator fire, and after they got down, they decided they could probably fly after all. Well, in the meantime the wingman had removed the pilot and the co-pilot and returned them to Marble Mountain to pick up another airplane so they could continue their mission. Well, as they went back out, they took me and another pilot back out to the aircraft and we turned it up and flew it back to Marble Mountain. But it was kind of interesting. We went out there and went running off the back of the airplane. The way the troops do. And saw all the grunts standing around watching us. Then we sort of wandered over to the plane, started it up and lifted out.

Well, last night I did something unusual for me. I sat down and wrote four letters. Three of which were four pages long. That is unusual for me, as I know you will agree.

Since the last time I spoke, we've moved into air conditioned hooches, which is pretty nice. They're smaller than the ones we were in before. Actually, it's a Quonset hut that's been partitioned into four separate rooms with an air conditioner on either end and an opening up on the ceiling between the two rooms that are side by side. We've got three guys living in each quarter so it works out to 12 guys living in the Quonset hut. They're a lot cleaner. We have metal wall lockers to keep our stuff in. Also, there's a marked decrease in the number of roaches that we see swimming around the area which makes life a bit more pleasant. All in all, we like our new situation a lot better. In fact, I am sleeping better now that I can sleep under a blanket.

Well, that mortar attack I spoke about before, it never happened. I thought you'd be glad to know.

July 12, 1968

Last night on the Bald Eagle we had to launch, first time we've had to launch in a couple of weeks. We were supposed to pick up 250 pounds of blood and take it up to Phu Bai. We got over, got the blood and then had to come back to Marble Mountain to get one of my radios replaced. Then after we got that done and had started to taxi out, we got a call from the controlling agency here at Marble Mountain, a place called Luminous Base, that we had to go pick up an emergency MedEvac.

Well, they gave us the brief on that one, and then mentioned something about another one in a non-secured zone. So we just told them a negative on that. We're not going to pick it up right now. So we went out and picked up the first one, a snake bite victim, brought him back to the hospital and headed on up to Phu Bai with the blood. That was pretty uneventful.

But the thing is, when you launch, you just never know. I guess I've really been pretty lucky so far. Most of the missions I've flown on were pretty good. I understand that today, I am to fly a USO run which sounds like fun. Well, it's about 20 minutes till 11 right now. I've gotta get dressed and get over to the mess for something to eat, cause I've gotta get on down to the Squadron. I've got a brief in about an hour for my hop this afternoon. So I'll talk to you later.

Well, here I am again, same day. I had a pretty good time on the USO run. We hauled two different troupes around different places. The first one was rather exciting because we hit a tree. Kind of made the people enjoy their trip a little bit more; at least it was something out of the ordinary. After we dropped them off, we had to go back and get another airplane because we broke the UHF antenna off the bottom and it also put a hole in the Plexiglas up in the front. Then, we picked up another load and hauled them out to another zone and dropped them off. Then we went back to the first place, shut down and listened to the show.

It was pretty good. A bit gross, but pretty good. Then after we picked them up, we went out and flew around a little bit while they shot the machine guns. They thought that was really something. Of course, I guess it is if you don't see it every day.

Well, I came back to Marble, got your package and brought it back to the room. Had to go to chow because we had a show here tonight. It was just like all the rest, at home no one would even bother to go. Then I came back to the hooch, opened your package and was really thrilled with all the goodies I got. And, I was kind of thinking, you know, a package like that once a month really helps the morale. My roommates and I really got a kick out of it. Particularly, tell Jim we liked the dart gun.

I was really glad to get that alarm clock because that is going to be pretty handy. It's the only alarm clock we have. My moustache scissors are real fine; they will work out well. My moustache is getting just about to the point that I can start trimming on it. It's a month old tomorrow by the way and…Oh yeah, tell Warren thanks for the fly swatter. That's going to come in pretty handy. We'll use it a lot. And thank him again for the dart gun.

Well, it's after 9 so I think I'm going to hang it up and get some sleep. I'll talk to you again.

Chapter 4

Settling In

Hi folks, John here. Tonight is the 13th of July, a Saturday night. A pretty unpleasant night. We had a soft rain this afternoon and I was supposed to be the night turn-up pilot. But as luck would have it, they set Condition 3, and they don't turn-up airplanes when a Condition has been set. So, here goes the old Ace skating again. Did I mention that that's my nickname now, Ace?

Anyway, I would like to say that when they start setting these damned Conditions, it makes the hackles start rising on my neck. I mean, I'm telling you, these rocket attacks don't do anything for my morale. We heard today that earlier this afternoon, they captured 47 – 107 millimeter rockets. So now word has it that sometime between now and the 17th they are going to really hit us.

It is a well-known fact that all the Vietnamese units in the Da Nang sector, now, there are at least two divisions of them about, are re-supplying and getting ready to move. So it looks like things will soon be picking up.

One good thing for our team, though, is that Da Nang itself is surrounded by a great deal of flat land. This makes it difficult for any troops to mass and make any organized movement toward the complexes here. With helicopters overflying on routine supply missions every day, you've got dozens of eyeballs scanning the area. It would be almost impossible for a conventional force like the NVA to mass and move against any of the compounds here without being discovered. The hills are too far away for that. Well, I might add some more later on. No telling what's going to happen around here; we'll talk to you later.

Well, it's later. I just got back from an aborted turn-up. I got called that they wanted a turn-up done so they sent the truck down for me and I got down there and I no sooner got there, than I was told my turn-up was cancelled, so I got the truck to bring me back. Upon my return, my roommate Tom W. informed me he had used some of my roach killer to kill a roach. I thought that was very interesting and asked Tom to share his dramatic tale with you so you would know how well the roach killer you sent over here is working out. Take it away, Tom.

"Thanks, John. Hi folks. Well, as you know, we all appreciate it here in the hooch when someone else gets one of these CARE packages. And as you know, we do have some awesome roaches tooling about here in our hooch. And as it happened, I saw this roach go by and I looked around for a weapon. I saw this can. The label read: Roach Trapping Powder – Odorless, Non-Staining. Net Weight One Pound. So, I reached for the can but I discovered there is no way to open it. I was a bit perplexed and the roach was getting away. So I decided the only thing to do was to use the roach trapping powder most effectively so I picked up the can and smashed the roach. And, I'm here to tell you it just does one hell of a job."

Well, that gives you an illustration that Marines are never at a loss when confronted with a new situation. Now you can see we have used that roach killer to full advantage and I am sure we will continue to do so. Well, that's all on that episode. More later.

July 14, 1968

Here I am in the coolness of my room listening to my friend, J.S. play the guitar. And my friend Bob W. accompanying him in song. Well, I got to sleep in late this morning, which is a little unusual, but it helps make Sunday a bit more like Sunday.

I went out this afternoon and flew some re-supply for Operation Pocahontas, which is a sub-operation of the overall Mameluke Thrust. I flew into an area where last night our guys took quite a bit of shit. We had to launch our Bald Eagle birds, our emergency MedEvac birds, and they really took some fire last night. They were catching 50 cals all the way up to 3,000 feet. I went into the zone, there were 19 MedEvacs to come out, and it was 50 cal fire all the way. All this without the Hueys, mind you. But we got all of 'em out and all.

We went out today and re-supplied some of the units in that area. But it was pretty quiet this time. It's kind of funny. You know, there's people down there, bad guys, but nothing happens. I don't know, I've been pretty lucky so far. No one has chosen to shoot at me. I had a pretty short day, only four and one-half hours and it wasn't too bad at all.

I just checked and I'm supposed to fly the same thing tomorrow, re-supply for Operation Pocahontas. We fly these re-supply missions every day. We haul water and ammunition, stuff like that, to the grunts in the field. Sometimes we take them goodies too, like beer and sodas. I told you once before I think about taking ice cream out. I haven't been getting many letters over here. I got one from you all, and am looking forward to a tape from you. I've gotten five or six letters from Anne. I got a letter from Mrs. S. and also one from Mrs. T. and, I got a couple from Mary.

By the way Mom, I wonder if when you get this tape you could go into the den and look up on the top of the bookshelves for me. There you will find a large blue book with nothing written on it and inside it, among other things, you will find a copy that Bob what's his name wrote out for me, the words to the song, "Plastic Jesus." I'd appreciate it if you would send that to me, so my friend Steve and I can memorize the words and sing it.

Well, things have been pretty slow. Hope tomorrow is the same. They did decide tonight to have a fly-away. A fly-away is when they take a lot of the birds and fly them away to some of the more secure zones in the area to get them dispersed in case we get hit by rockets. I lucked out, wasn't on any of the fly-by crews.

I did discover something new today when I was out in one zone. We shut down to eat dinner because we were flying at supper time and there were some artillery pieces firing out there. I never realized it, but you can actually hear the projectile as it passes overhead. You can hear the gun fire, and then it's a sound kind of like when you hear lightning crackle, that's the sound it makes passing overhead. Imagine how bad it would be to be on the receiving end of some of this stuff.

Oh yeah, like I said, the CARE package was well received. It boosted all of our morale. It's kind of funny, but anytime somebody gets a package like that, one with a lot of buried things in it, guys in the room get involved. It's as much fun for them as the guy getting it. So I'll tell you one thing, some chocolate chip cookies would be awful good. I'm getting thumbs up from all the guys. Chocolate chip cookies would be really great. My friend Tom W. had a problem. He sent a check home the first of July and he just heard from his wife that it never got there. He went to the Post Office and they said they were sorry but there was nothing they could do. And he went to disbursing and they told him the same thing. If they stop payment on the check, it will be six months before he can get reimbursed. In the meantime, his wife is stuck with all the bills, which as you can imagine, really sits well with Tom. He is pretty hacked off at the Marine Corps. As he put it, well, I'd better not say how he put it. He's pretty much down on the world right now.

That's it for now.

July 16, 1968

Last night I was on the Bald Eagle again. I had a 15-minute standby alert for emergency MedEvac shit sandwiches as we refer to it. I'd hoped that we wouldn't launch. I was scheduled to fly with one of the Majors who is something of a screamer. But we did. Went out on the first one, a pretty high zone, 4,800 feet, which is sort of marginal for us in that we lose single engine capability and the ability to hover off of ground effect.

Well, we headed out there and we ran into quite a bit of clouds. In fact, we flew blind for a while. Finally, we got up above the zone. It was marked by four strobe lights and six flash lights. Nice. Well, the good Major I was flying with, a delightful man except that he leaves his patience on the ground when he goes flying, was trying to get the brief on the zone and had a bit of trouble getting the answers he wanted from the people on the ground. As far as I could tell, he was getting the answers but he was just confusing the issue for everybody. As a result, there was a failure to communicate between us and the ground.

While all this was going on, it clouded up and there was no chance at all of us getting down there. So we finally just had to cancel it. We started back to Marble Mountain but then we got re-fragged. By re-fragged, I mean we got another frag order to go pick up another MedEvac. Well, this other one was out in bad guy country. Deep in bad guy country. And two nights before, whenever we had to launch a Bald Eagle down into that area, the birds took a lot of fire. So needless to say, my old pucker string was pulled up pretty tight.

But we went out and didn't take any fire or anything, though on the way out, the Major was flying at 1,000 feet. That didn't help my pucker factor any because that is well within small arms fire from the ground. But I guess the Major knew what he was doing because we got out okay.

We got back to Marble Mountain and there was a hospital right across the road from the air facility, so we dropped the wounded guy off there, shut down and spent the night sleeping in the ready room.

I got up this morning and was supposed to have a truck at my disposal to make a run for the coffee mess. I am Assistant Coffee Mess Officer which is really a cherry job. But there had been a screw-up in the motor pool, and they didn't have a truck available. So I've spent the day trying to scare up a truck without any luck. I may have one lined up this afternoon, I'm not sure yet. I'm gonna go out and buy some sodas to sell in the coffee mess.

I got your package yesterday with all the batteries and goodies in it, and now I probably have more batteries than the exchange does. Not only did I get the 12 you sent, but I got five in the machine that I bought, plus another seven I had picked up in the PX just to make sure I didn't run out. So altogether I must have 30 or 40 batteries now.

Well, I just plan to lay back and enjoy this day as much as I can. Days you don't have to fly are good days because you don't have to worry. I came over here with the idea I would rack up all kinds of hours, that the guys who didn't fly a lot were jerks. Then I got over here and decided I wouldn't try to rack up lots of hours after all. I'll fly when they schedule me but when I am not scheduled, I'll be hard to find. As I am going to be hard to find today. So maybe today I'll come with some good things to put on this tape for you. I'm hoping that maybe by this afternoon that first tape from you will come. You mentioned it in the letter I got yesterday. Oh, yeah, by the way, thanks for the pictures. They came out pretty well and I enjoyed them.

So I'll talk to you later.

July 17, 1968

I just finished listening to your tape, and there's a few things I want to say before I go to sleep. First, I want to clear up the thing with the watches. I mentioned I was sending them home; I meant home to the States. Two of them went to New River and one to West Virginia. They were for friends who wanted watches, so they won't be coming home. I've already received acknowledgement they got to their destinations. I kept one of them and the guy who was with me kept one. So all five watches have been accounted for there is no problem.

I enjoyed the tape so much I'll probably play it a few times. Tell Howard next time you see him that I recognized his voice right off, that I do remember the night we all got together, and I'm looking forward to some more of that when I get home. Tell him also I ran into a guy from West Virginia in the

Squadron today. Apparently, there's a couple more of them, one from Huntington, and they have a West Virginia flag they will present to me when they leave. I will get some pictures of me with it to send to you.

You can tell Uncle Joe I was completely overcome by the words of wisdom he put on the tape, and I was going to send him a letter, but I didn't know his address. Tell him I'm not going to try to end the war over here because I'm not greedy. I'd like to save some of it for all those other people who are dying to get over here. Tell him not to worry, I'm coming home to pay him back that steak dinner. I try to get out of some things, but I won't try to get out of that because the consequences of trying to get out of it would be far worse than paying for the dinner.

Tell Pop I enjoyed his singing and that I'm going to play that for the rest of the guys that I live with tomorrow. I'm sure they'll enjoy it too. Well, I had a pretty easy day today. I had an administrative run down to Chu Lai. I made one run in the morning and one more in the afternoon. Carried some gear and some passengers down there. It's a pretty safe run, no problems because you fly over water. Don't have to worry about getting shot at. The guy I flew with was a real nice guy to fly with and we split the time up very well, about half and half. All in all a nice day.

Well, I better close out now and get some sleep in case I have to get up early and go to work. I'll check in with you tomorrow.

July 18, 1968

The sounds you just heard were the sounds of a hop I flew yesterday. I was in aircraft 51, a run down to Chu Lai. Today was pretty slack. Went out and fired my weapon this morning. Didn't get up till about 9 a.m. All of us didn't fly this morning. In fact, Tom and I didn't fly at all. Then I went down and did a little work in the coffee mess. You know, it's kind of ironic. They made me Squadron Coffee Mess Officer and the job entails keeping books and running an operation like a small store. If they had any idea of my record in financial dealings, they would have never given me this job.

This evening, Tom and I went on a picture taking spree. We got some really good shots. We went down to one of the old-fashioned heads and each of us went in and had a friend take photos of us sitting there. These should be classic. Then they also had some old-fashioned urinals around back and we got photos taken of us using them. They should be great.

I had to go pre-flight an airplane because I'm on the second section of the Bald Eagle tonight. So Tom went with me and we got some photos of us acting like real pilots, getting our combat gear on, and also photos of the 50 cals like we were firing them. We called it Hollywood Night on the Line.

Also, I could also use some moustache wax so I can start training my moustache. Tom's at my elbow here, and he wants to add a few words:

"Ace was kind enough to let us listen to the tape that he got from you and all the folks in Barboursville, and I was just lying here thinking I'm from a small town, too. I'm from D.C., and we have the same type people. You know, people are the same as far as I'm concerned the world over and you run into the same types of schoolteachers and policemen. It was really nice listening to it and humorous to someone who is not from Barboursville. We sure got a kick out of listening to it.

It's difficult for people back in the States to understand the situation over here. We're not in danger 24 hours a day as you might imagine. We really do appreciate the thoughts and the letters and all the things you people send to us, even though it maybe doesn't always seem like we do."

Well, I guess with those few words of Tom's there, I'll close out for tonight and perhaps tomorrow, I'll come up with something new to say.

July 19, 1968

Today marks one month in the country for me. We were just sitting around discussing that there are a lot of small items we take for granted, things we never thought too important back in the States. Over here their value really gets out of proportion. For instance, we carry this little knee board around that has a 15 cent pad of paper mounted on it. The whole world stops when we run out of these little pads we can't write down mission credits or notes to our families, or anything. We wouldn't know how many air medals we got, the entire system would collapse. Or red lens flashlights. Without these little red lenses in our flashlights, we couldn't go anywhere at night. And the same is true of pencils. What would we do without pencils?

And I can't help but wonder what do the VC do? How do they get by? Does someone just hand them a box of rice, a box of Uncle Ben's? When we go to the Officers' Club and there's no ice, we go nuts. No ice in our drinks! My God, you should hear the noise. Do you suppose the VC get ice in their drinks out there in the bush? Do you suppose they get drinks?

That poor VC. That poor asshole puts a rocket in his back pocket and walks 100 miles to a battle zone, shoots it, and then has to go back and get another one. They can't fly that stuff in. The VC doesn't have an air force; we would have noticed it by now if they did.

Well, we were just talking about some of the officers we have to fly with around here, some people who don't exactly inspire confidence, if you know what I mean. The other night I flew into a zone with this one Major and I noticed when we climbed back out there was a glow around the airplane. I said, Major, is our searchlight on? He says, now, it isn't but if I hadn't said something about it, he never would have noticed. He'd of left the son of a bitch on for all the world to aim at.

I was at New River when every one of these Majors that we have now in our Squadron was transitioning. There was this one guy named Phil J., he was one of the instructor pilots down there at 261, and he was working with these Majors and I used to run around with him. He told me once in all sincerity, John, I don't envy you a bit. Because you're going to go over there and these Majors are going to go over there, and they're going to be HACs. You're going to be flying with these guys, and they're going to get you killed. And he was dead serious about it. And sure enough his prophecy has come true because these goddamn people are over here, pulling the same shit they pulled at New River.

And these guys are supposed to be our leaders. I remember this one time we went into a real shitty zone, I'm flying the airplane and this fool decides he doesn't like something I'm doing, and he grabs the stick right out of my hand. I screamed at him to get off the goddamn controls. Don't ever do that to me again. You know, if I'm in a shit sandwich, I want to feel everything the aircraft does, which you can't do if someone is messing with the controls.

Well, as you can probably tell, I'm not feeling too upbeat right now. So I'll just shut down and try you again when I'm in a better from of mind.

Chapter 5
Serious Business

Hi folks. Today is Sunday, 21 July. Well, it's about 11p.m. now and I just got back from one of the biggest shit sandwiches that I have ever seen in my entire life. I'd heard about things like that, but I'd never seen one until today.

The day began innocently enough. I was supposed to fly at noon, but the mission was cancelled. So I went into my coffee mess, the little business I manage. Another laugh. I was sitting in there working quietly when they came in and said: "You're going to have to fly."

So I went into the ready room and was informed by the Operations Duty Officer that I was going up to fly in a shit sandwich. I just laughed at him, but little did I know how right he was.

Well, I went out and pre-flighted the first airplane they had given me, and we had to down it for a minor mechanical thing. So we got another bird and lifted out for An Hoa, which I think I have mentioned before. Got out there and shut down, got a briefing. Seems as though earlier in the day some of my fellow pilots had put in a reaction team. Now, a reaction team is a bunch of guys sent out when someone reports z*ps running around on the ground. Some General had this idea it'll win him a promotion, I guess, but we launch some birds and get some people out there quick to try to kill the z*ps. Well, after a while someone has to go out and extract these people. Only this time they hadn't killed all the z*ps, and we had to go get them anyway.

So we got a brief. Everything was still quiet, the sun was shining and all, a nice quiet Sunday afternoon. Somewhat like the ones you have at home. We started up the birds and flew out to this place about three miles north of An Hoa. An Hoa itself lies in a large, flat valley with mountains to the north and south. The valley narrows to the west, and rises back up into the mountains. So we took off and headed in. When we got there, we found four Hueys orbiting over the zone, armed with rockets and machine guns.

We went in at 2,200 feet. The leader of our four-plane division took a hit. We were doing 110 knots. The round went through one of his hydraulic lines, so he had to go back immediately to An Hoa. We talk about getting shot down. Well, this guy was shot down, and by that we mean he took a hit bad enough that it forced him to go down just as fast as he could get down safely on the ground. So we followed him back to An Hoa to make sure he got there, and then went back to the zone.

In the meantime, the first two birds had gone into the zone and taken hits, one a them in the cockpit. Well, the fear factor began to rise and the pucker string on this old asshole of mine cinched up several notches. But there was worse to come. We decided to go back to An Hoa and wait a while until the Hueys could call in some Fixed Wing to work the place over. While all this was going on, we called two more birds to come out.

So now there were four of us again, and we went back out to extract our people. Well by now my pucker string was almost breaking. I knew this was one of those shit sandwiches people are always talking about, because there was all kinds of firing going on down there. So we went back out and orbited a bit north of the zone waiting for the Fixed Wing to go in and run some air strikes.

Meanwhile, the Hueys had gone back, refueled and loaded up on rockets, and come back out. They were making gun runs around the area and taking fire from 360 degrees every direction. Once again we were the fourth plane in the division. The first two birds went down. Immediately, when they got down below 1,500 feet they started taking hits. They went down into the zone, picked up their people, and lifted out. They took fire going in and they took fire coming out. And all the while we're orbiting up there, I'm watching this shit, and I tell you I have never seen anything like that in my life. And the part that really scared me was knowing I was going to have to go down there right behind them.

Well, the first two birds lifted out and our leader called to go down, so we started following him down. We got down to about 1,000 feet, and our leader lost the zone. At this point, my pucker sting about broke. Here we were flying around at about 1,000 feet and everybody in the whole damned world is shooting at us and I should say that 1,000 feet is well within small arms fire and to be honest with you, I was real scared.

So finally everybody was screaming for our people on the ground to pop smoke so we could find 'em. They did finally and we sat down and we could see tracer rounds all over the place. The Fixed Wing were running strikes all around us, not more than 100 meters away. The Hueys were making rocket and gun runs to the z*ps who were shooting, the friendlies were shooting, our gunners were shooting, everyone was shooting.

Well, when we got down, we found out they only wanted one bird to come in. They weren't going to take two loads out then because they had just enough people so that two birds could come in later and take everyone out at once. So there we were sitting in that mess with no one to take out. So finally our leader lifted out and we lifted behind him. He took a couple more hits while he was climbing, but up that point, my luck still held out. We still hadn't taken a hit.

We were going back to An Hoa, three miles to the south, at about 2,200 feet at 11 knots and all of a sudden we heard a big bang. Well, I thought our gunner had fired. It sounded like a 50 cal firing one round, but the pilot looked over at me and said: "we took a hit, we took a hit. It's over there beside you, ok, look!" So I looked over but couldn't see anything. So I looked closer and still couldn't see anything. But when we got on the ground and shut down, and checked for battle damage, sure enough, right below my seat, in the side of the aircraft, was a bullet hole. Yessir, my first one!

Well, the other two birds had to go back in and extract the last people, so all told we had six birds in on the action, and every last one of them took hits. We were the most fortunate because we only got hit once. The others had three, four, five hits, and one had six. But our bird and one other were the only ones to get hit in the cockpit. The other one took a round in the window canopy, completely shattered. The co-pilot is a friend of mine and you should have seen him; his eyes were big as grapefruits.

Well, I think it's fair to say our Squadron got the living shit shot out of it, because we have nine birds down tonight with battle damage, and only two ready for duty, and they need work. But we were fortunate in that nobody got hurt seriously. One crew chief took a round in the leg, and one of the wounded was hit when we were pulling out.

I did learn one thing today. You may recall me saying in the past that when I got in a situation like this I would be too busy to be scared. Well, folks, I'll let you in on a little secret; you're never too busy to be scared. I've never been quite so scared in my whole life. All I can say is I'm back in one piece. But my luck didn't hold, I was in the country for a month and two days when I finally got hit. Well, that's it for today.

Oh yeah, one more thing, that chicken shit stuff started again. My roommate failed to salute one of the Senior Captains in our squad today, so all the co-pilots got called in. I'm sorry I'm talking so low, the guy I'm talking about lives on the other side of the wall here. I missed it because I was busy, but the others got a lecture on military bearing. You should always salute Captains, call them by their rank, even when you're flying and you should always shit on the enlisted people every chance you get, and so forth and so on. Now every other Captain in the Squadron says you can call him whatever you please, they don't care, but this one guy is a stickler for that sort of thing. So I guess for a while, until it all dies down, we're gonna have to be garrison soldiers. Well, that's it for today.

July 22, 1968

Hi, it's me again. Then again, who else would it be? Today was rather quiet. I was supposed to fly but it was canceled, which was all right with me. I need time to get the charley horse out of my derriere from the clutch situation I was in yesterday. Oh yeah, a little epitaph to that. We had to go back in today to extract two people who had been left in the zone. Isn't that great?! They left two people in there and so they had to send a whole raft of airplanes back in today. But it all went smooth this time and they got in and out with no problems.

This morning the bird I was in yesterday was in the hangar with the other five birds from our flight. They had mine for some shop metal work. They're going to patch the bullet hole and then they had to find the bullet, because they didn't know how much damage might have been done to the controls. Well, they dug out the bullet and presented it to me. It wasn't a ceremony or anything. I was sort of bugging them for it, actually. And now I'm the proud possessor of a bullet that actually hit an airplane I was in, in Vietnam. I'd like to say that will be the last such bullet, but I can't. Maybe one of these days I'll pull a 50 cal out of my plane.

Well, other than that, my day's been slow. I spent the day knocking around my coffee mess. I'm pretty lucky in that no one keeps track of me, mainly because no one much comes over here. I spent the day selling sodas, had a rather good day - sold close to $100 worth.

Well, I didn't have any adventures today, but my friend Tom did. So I'll let him tell you his story. Take it away, Tom.

"Hi folks! Well, Ace asked me to describe my hop today and I guess it was pretty typical except that we took an unusual number of rounds. But it wasn't quite as exciting as Ace's last night because none of them came through the cockpit. Our job was to insert some recon teams. These teams usually consist of about 10 people. The first zone we went into we had to extract some people. We had to lift up a recon team that had been in the field for about three days.

Well, this is fine and dandy, but they had their secure zone which meant they had security out probably 10 meters in each direction, and they had had contact with the enemy four hours before. Well, we said fine, we liked the mission brief. Our Hueys went in to clear the zone for us. But as soon as we sat down in the zone, we took fire. Our immediate reaction was to call in the coordinates to get some Fixed Wing in to do something. They did nothing. So we lifted them out, returned them to the secure area, and went back out to insert the fresh team.

Well, this proved somewhat tricky. The recon people chose a zone right at the bottom of this big cliff because they had found some caves there. Well, okay, but the zone at the bottom of the cliff left us vulnerable to falling rocks, or anything else Charley might like to throw at us. And sure enough, no sooner did we put the recon team into the zone than all hell broke loose. I never heard so much firing in my life. They were shooting at us from all directions. The recon team returned the fire, but there was no point in us shooting back because the enemy was entrenched in the caves up above us. The Fixed Wing dropped a lot of ordnance into the zone, but left the caves untouched. Made no sense at all.

Ace and I have been talking and we sure can't figure out why they do some of the things they do. We're confused, as usual. We just fly our hops and try to get enough flight time in. But we just hope the people who make policy will catch on to some of the things we see happening all around us. It's hard to convince yourself that you're doing something worthwhile when you've seen so many things go wrong. Well, we're paid to follow orders, and we do. But we fly with a heavy heart. There's no more we can do for the poor grunts we're supposed to support."

Well, Tom hasn't had time to get his thoughts collected, as you can probably tell. He's still a little nervous from his experience this afternoon. Although I don't blame him one bit. It's really tough to have to go back out there when you know damned well before you even get started what's going to be waiting for you. Like Tom said, we only wish some of these policy makers could go out and fly on some of these missions they send us on. I'm sure there would be a drastic change in the way this is being handled.

Next time you send one of your CARE packages, I sure would appreciate some moustache wax. My moustache is really unruly, and I have trouble managing it.

Also, I'd appreciate it if you would send along the tape Jim T. and I made while we were driving around Barboursville; these guys would get a big kick out of it.

Well, that about gets it for today. Tomorrow I'm to be SDO, Squadron Duty Officer. Which kind of makes you Squadron boy. You just sort of go running around doing useless things like inspected the enlisted hooch area, chasing people, trying to locate people, waking people up, that sort of thing. The good thing about it is I don't have to fly.

July 24, 1968

I guess the last time I talked to you was Monday night. Tuesday I had the SDO. Monday night was kind of nice. We had a mortar attack here. Took about 40 plus mortars down around the flight line. Damaged two of our birds. Had one round go through the roof of one of the hangars down there. Pretty badly damaged one of the CH 53s. Other than that, it wasn't too bad. In fact, my roommates and I slept through it until near the end when Tom finally woke us up yelling: "Incoming, Incoming." So we all got up and crawled toward the door. Tom opened the door and said: "okay, it's safe" and we all got up and crawled toward the door. Tom opened the door and said: "Okay, It's safe," and we all dashed for the bunker. Only as we got there everyone was coming out because the all clear had sounded. Just goes to show ya.

Tuesday, when I was SDO, they put us on a Condition 3 which is an attack probable category. Everybody has to carry their weapons and wear their flak jackets, helmet and the rest of the garbage. But nothing came of it. Then they put us on Condition 3 last night and again tonight. A while ago we got word Da Nang is to take some rockets, and maybe us too, I don't know.

Today I went over and made another beer run for the coffee mess. The enlisted guy I've got working for me, a really sharp guy, went up to the Vietnamese who was guarding the beer and asked for a couple of cases. The guard gave him a very definite negative on that. So while they were loading beer on the truck, my man calmly walked into the tent, put a case under each arm, and calmly walked out. He got back to the coffee mess with slightly more beer than he actually purchased.

Now, I know this is unethical by the standards back home, but by the standards out here, it is perfectly all right if you can get away with it. When we got back, my man Tom had somehow managed to borrow a jeep. So we went over to the Army compound next to our base which is a main supply depot for the Army troops in this area. Tom has a pal over there. So we completely loaded the back of the jeep with all kinds of gear – boots, skivvies, shirts, towels, and socks and came back here to open up shop. We don't sell the gear, just pass it out to our friends. Next week I understand they're getting in 200 and some odd rain suits over there, so we're already taking orders for 'em.

It's kind of funny, but the Marine Corps supply system isn't quite up to getting all the stuff we need, so we've sort of turned to scrounging, although I guess by the standards back home it would amount to thievery. Over here the situation is different and warrants different morals. We figure as long as we're only getting the stuff we need, it's okay.

Oh, by the way, I got the cookies today and let me tell you they were really popular. In fact the can is almost gone already. Everybody who got in on it said they were great, so feel free to send some more. Rest assured they are appreciated to the fullest.

We had a show tonight over at the Officers' Club, pretty decent. Some Koreans. Had three girls and a dance band to back them up. Tom and I were discussing how much better these Oriental girls are beginning to look to us in the last month or so. We're kind of looking forward to our first R&R in Bangkok. Well, I only got about two hours sleep last night and I haven't had a chance to sleep today. So I'll close out now and talk to you later.

July 26, 1968

I was just thinking tonight is Friday and all the bachelors back in the States are out chasing women or drinking and things like that, and there I was in that fabulous fun spot of the Far East in another shit sandwich. We were out today on a regular old re-supply mission when all of a sudden we got word we were going to be diverted to pick up 50 emergency MedEvacs. They've got this deal where you can divert aircraft to do this stuff and they got us over there in a real mess. It took about 40 minutes to get everybody organized. Fortunately, I was the last bird in the last section and got to watch the whole thing from 5,000 feet. They got in and discovered that instead of 50 MedEvacs they only had 30. The ground troops had been mortared. As soon as we got the wounded out, it was decided we had to go back with water and ammunition so the grunts could last the night. So I went back and this time they had two more transport helicopters in addition to four Huey gunships, and four Fixed Wing doing air strikes. One of our birds got hit, but nobody was hurt. A round came through the back of the airplane, went through the controls closet, missed everything, and lodged in the armored seat. Those armored seats are a good deal.

Well, once again we were the last bird in the whole thing, and the only thing left was a load of chow. Well, by that time it was dark and chow doesn't rate an emergency after dark. So we didn't have to pick up the load and take it to the zone. So we climbed out of there and went up to 5,000 feet to watch. And there was some shit going on out there tonight. There was intense automatic small arms fire coming up at our birds that were orbiting around making these fantastic rocket and machine gun runs. At night you see all the tracers; oh, it is really something to watch. But they got everything into the zone, shot up the countryside until we were out of ammo, and flew back to Marble Mountain.

Well, we got back and had chow in the enlisted mess because the officers' mess had already closed. Got back to the hootch and were sitting around when the siren blows for a Condition 1, which means we have two minutes warning. Apparently they were really afraid of that every night this week, so I don't know what to think.

Well, I'm to fly recon tomorrow, which means a 4 a.m. wakeup, and it's after 11 now. So I'll talk to you tomorrow when I get back.

July 27, 1968

Well, I made it like a real civilian tonight. Took a shower, put on my Levis and loafers, and just laid on my rack reading magazines. Kind of good.

Well, this morning at 4 a.m. as predicted the SDO came in, woke me up, and I said, don't even tell me. I know already. I got up, got dressed, didn't even shave, just brushed my teeth, went over to the recon pad, got the brief, and it didn't look all that good. The Section Leader said we probably couldn't find a hairier recon mission if we tried.

But it actually turned out fairly well. It was a long day, but pleasant. We put a couple of recon teams into zones. In a couple of them we took some fire and had to call in Fixed Wing, but all in all it wasn't so bad a day. I got 13 mission credits.

I was just thinking how much I missed my Corvette, how much I'd like to go tooling around Barboursville in it, doing wheelies and such, as in days of yore. And I'd like it if you'd send me the photos from all those rolls of film I sent you. You can keep the negatives and all; I'd just like to see some of the pictures, see how they turned out.

Well, today the pay roster came around, and you sign up for how much you want. I had $410 on the books, so I signed up to get paid $400 of that. I got this money so I could begin to look around for some good stereo equipment over here. I've been allowing myself about $10 per week, and it's been working out pretty well. And don't forget that tape I asked you for, the one Jim and I made.

Well, last night I went to make a head call and ran into the guy who makes up the schedule, and saw I had that recon mission. And right after that the siren blew because we were supposed to have a mortar attack, but nothing came of it. A bit more on the recon mission. We took fire in two zones, but I was in the second bird and didn't get into it. The zones we went into were perfectly calm, no problems at all. The other guys encountered some fire and had to call in the Hueys and Fixed Wing, which just pasted the hell out of them. They put in napalm, 20mm cannon fire and some rockets and frag bombs. Then the Hueys went over again. And the guys on the ground who shot back made a bad mistake. That's one thing about the VC, you can tell them from the NVA by the way they operate. Had they been NVA, they would have waited until a helicopter was in the zone and then opened up. But the VC get trigger happy and open up as soon as you come into view. Which makes it pretty difficult for them to get hits on you because you stretch your airspeed up and it's pretty tough to draw a bead on you.

So it was probably the VC we tangled with today. It's kind of funny, but you never know when you go out in the morning just how things will turn out. Some days will be nice and calm, routine flying. Others you think it's a routine re-supply mission when all hell breaks loose and you find yourself in the middle of another shit sandwich like we did the other day. Well, actually, twice this week, Sunday and then again yesterday afternoon.

I mentioned earlier that I thought I was getting adjusted to this thing, but I've decided now that I'm not. I just feel kind of listless tonight, you know. I don't know. I've decided that my mother just didn't raise me to do this kind of stuff. And it's kind of funny, but all the other guys I've talked to have pretty much decided their mothers didn't raise them for this either.

And still, here we are doing it. One thing that I wanted to say the other night, a feeling that I got that I hadn't experienced before, and that was when we were up there watching when the guys were pulling out those MedEvacs, and later on when they went back on emergency re-supply, and they were taking all that ground fire. I suddenly got the urge that I just wanted to go in there and destroy those people. I mean, I just wanted to watch them meet their end and to have a part in it. I wanted to put my finger on the trigger, but there's nothing you can do really but just hunker down in your armored seat and hope for the best. Of course we have a gunner in the back with those 50 caliber machine guns. He's having a heyday. Those kids get on those guns and just blaze like hell. They really love it. I told them once they should have joined the infantry if they love it so much. But they assured me they didn't join the infantry for a reason, and I kind of laughed and agreed with them.

Well, tonight was roast beef night over at the club, or the mess, where I eat. It was kind of good. I had two huge pieces of medium rare roast beef that in the States would have cost me $5 or $6. I had it for free. Last night I had a whole plateful of shrimp and that was all I ate, just shrimp. I was thinking that if I bought that back in the States, it would have cost $10 or more for all that shrimp. But here it's free.

Well, that's about it. I notice this tape is about finished, so I'll send it off to you. I did manage to buy two more tapes the other day, so I won't miss a beat. See you later.

Chapter 6
A Red Badge of Courage

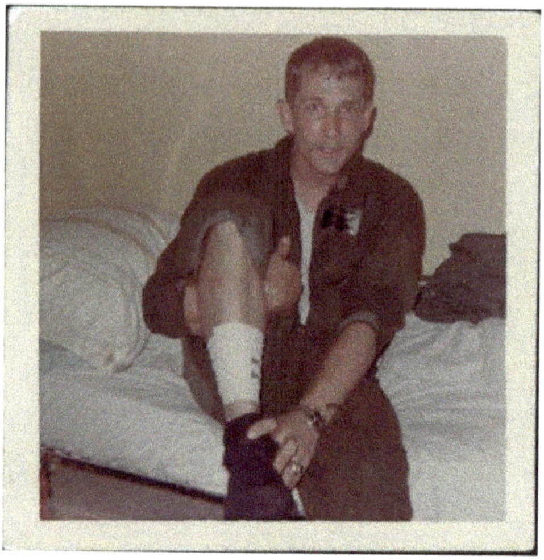

Hi folks, it's me again. This is the 28th of July, another good Sunday, and I got a story that's going to curl your hair. I went down and looked at the schedule this morning and saw I wasn't on it, so I decided to cut myself some slack. I stayed in the rack until 11 a.m. Then I got up, had a leisurely lunch, and all of a sudden the SDO came running in and said: "Come on, get your flight gear, you're gonna fly."

So I said, well, shit. But I got on my flight gear anyway and went down, and was supposed to launch at 1300 hours, but they didn't have any aircraft available. So we piddled around till about, well, it was 6:30 before we launched. We headed out on a routine re-supply, two birds and one Huey gun bird. Got our brief, picked up two externals, and started towards the zone.

Well, we got out there and the Huey gun bird went in and sort of reconnoitered the area. We got the brief from the people down in the zone. They said they hadn't taken any fire for four hours. Of course, when we hear that, we say: Are you shitting me? They're just sitting out there waiting for us. We were right. So the leader starts down into the zone, got down okay, dropped his load, no problem. He called that he was coming out so my HAC called that we were going in. And folks, all of sudden one of those fastest things in the world zipped through our cockpit, came in down below me, came up, cut a hole in my flight suit, cut one of my shoestrings in half, went through a first aid bandage I had in my pocket, zipped up, ripped the lip mic off the helmet of the HAC, and went out through the window up high behind his head.

Well, folks, immediately the old pucker string just broke because bullets banging around in the cockpit are not much fun. Well, another thing was my kneeboard flew in half. The top half flew up and the bottom half was still strapped to my leg for a couple of seconds there my mind didn't even function. Things like that happen so fast you don't even think. Then all of a sudden I looked over at the HAC and he was gesturing wildly at me, pointing at his mouth, and I thought…Oh my God, he's been hit!

Well I didn't see any blood, and then I noticed that he didn't have a mic anymore. Then I understood his mic had been shot away, so he immediately waved off the approach. We were at 1,700 feet. I got on the radio and of course my voice had gone up about 10 octaves. I sounded like a young lady. But I called the other bird and told him that we were taking fire, that we'd been hit. So we started flying back to the zone where we'd picked up the gear. About that time the crew chief called us up and said: "Sir, we've had a fuel line cut." And then I really got worried because I could see us all turning into a big ball of fire.

Well, the HAC kept his cool and apparently I kept mine too because none of us panicked. We got back into the zone, dropped the externals, sat down immediately, shut the aircraft down. Now on the way back I noticed there was a numbness in my leg. And after we shut down, there was still numbness in my leg, and I was afraid to look down and see just exactly what was causing it. So I climbed out of the cockpit and got outside. I looked at the leg of my flight suit, and I looked down there and well, you guessed it, there were two holes in my leg. Well, they weren't bad ones. They were just little bitty holes, but they were holes all the same. They were on the shin of my right leg. Well, we had crash trucks standing by. They came out and began hosing down all the fuel that was everywhere, and they had a truck come out and pick me up and take me over to sick bay. They looked at my wound, put a bandage on it, and let me go back out to the zone. We picked up our gear, got on the other bird and they brought us back.

Well, immediately when I got back it was just amazing, everybody was out there. They'd heard the Ace had been shot. It's one good way to find out who your friends are, I reckon. They all come out to see ya. Then I went to get my leg x-rayed. I got two small pieces of shrapnel still in my leg here. They wanted to cut them out, but I didn't want those guys cutting on me. So I told them to forget that shit, just leave it in there. So they gave me a tetanus shot and bandaged it back up.

I got something to eat and went on down to the club. Well, a bunch of the guys from the Squadron were in there, and so they hailed me over and started buying me drinks. That was a lot of fun. But I'll tell you one thing, those z*ps have gone and pissed me off now. As I said earlier, I just wish I could go down there and kill every one of them. One good thing, we had Fixed Wing on station. I don't know why they were called out, but as soon as they could talk to us and find out where we took the fire, they ran some air strikes in the area. I just hope they killed those bastards. Guys that have the nerve to sit there and shoot at us. And it's kind of funny, well I said that. I guess I'm talking in circles, but I broke my pucker string again tonight. That must make about four of 'em now, so you may have to send me some replacements pretty quick. And, well, I've got a Purple Heart now, so it's another medal for the Ace. You can call around the thriving megalopolis of Barboursville and tell everybody, if you like, that your son has done his part for the war effort and is now the proud possessor of a Purple Heart. I just hope I don't get any more.

Well, I'd like to finish this tape tonight and get it off to you. I've got two tapes from you. I haven't had a chance to listen to them yet, but I'm gonna do that as soon as I finish this. I did get a couple of pictures of all this that I'll send to you. So you can see how minor it really was. I don't want you to get upset about it. I wouldn't be here dictating this tape if it had been serious at all. It's just that I was really lucky today, I mean really lucky. I just hope my luck continues. All the guys have been kidding me and telling me I've got a magnet ass, because this is two Sundays in a row we've taken rounds in the cockpit. I just don't know what to say.

Well, I'll say two things that I wanted to find out, things I wanted to learn when I came over here. One of them is what I would do in a stress situation, whether I would keep my cool and get the job done without going to pieces. Well, tonight I found out. As we say in the Marine Corps, I've got the balls to do the job. I didn't lose my cool. We got the airplane back, I did my part and all. In fact, I had to do several additional things that a co-pilot doesn't usually handle because of the fact that the HAC had his lip mic shot away, but I managed to get everything done and we got back, shut down the aircraft according to proper procedure.

One thing that was kind of funny. Normally when we shut down, as soon as we pull the engines off we've got the auxiliary power unit still running. You're using it after you get the engines shut down to build up the hydraulic pressures in the system so that they'll be there when you get ready to start your auxiliary power unit again the next time, because it's started by hydraulic pressure. Well, every time you shut down, after you get your engines secured and everything, you're on the ICS and you call up the crew chief and ask him if the pressures are up. As soon as the pressures come up, he'll tell you and then you can secure the auxiliary power unit, or the APP as we call it, it's the Auxiliary Power Plant, but anyway you shut that down and then secure the battery and everything.

Well, today when we got into the zone, there we were, fuel all over the place, crash truck at the ready, me with two holes in my leg and both me and the pilot in a state of high excitement. And of course I was taking care of everything because the pilot had his mic shot away and couldn't talk to anyone. So as soon as we got the engines shut down, I called the crew chief and asked if the pressures were up. And all of a sudden I realized what a silly question that was in view of the overall situation. So then I came back on and said: Fuck the pressures, let's get the hell out of here. So we shut down the aircraft and exited very quickly.

The other thing I wondered was what it would be like to get hit, and now I know. It's not painful at all. The two pieces of shrapnel hit my leg and it was numb. There was no feeling around the wound at all. It wasn't until later when my adrenalin flow eased up that I began to feel some pain from it.

Well, today's been slow and I've spent most of it in the rack. This morning I went down and the doctor sewed up the incisions in my leg, and after the Novocain wore off it started to pain me a good deal. In fact it was paining me so bad I decided I wasn't even going to walk to the mess hall to eat. But some of my friends saw my plight and brought some food back for me to eat in bed. Then one of them brought in some vino, a bit of wine, and we sat in here and had a bull session. Guys have been dropping by all evening. I guess I'm going to live now. I should be back in the saddle in about two or three more days, though I still have one piece of shrapnel in my leg that they weren't able to find. So I guess that'll be there for me to tell my grandchildren about. Who knows? By the time I get home, maybe I'll be able to tell the weather with it.

But the wounds will make for nice stories. I told the doctor not to sew them up too tight so they'll leave nice scars. So I'll be able to show my war wounds to people, which will be something of a bonus.

Well, I can't think of much else to say. We're not very talkative tonight. Unless Bob wants to say something. You want to say something Bob? Hang on, Bob has a word for you:

"Well, we've all been ribbing old Ace here about the kind of language he uses in these tapes he sends home. We're all impressed that you accept them, which is very nice on your part. But you know, considering everything, maybe it's a darn good thing that he does talk in the vernacular, because for one thing, it is the vernacular of military life, and for another, maybe it's the best way to describe the emotions you have in the area we happen to be in. So maybe Ace's dad can accept it better if he thinks of it in those terms. It's part of our environment, so I guess you're gonna have to put up with it, our awful language, for quite a while. I thank you and goodnight. I'll turn you over to Ace again."

Well, folks that was my roommate Bob W., better known as the Winesap, with a few words on profanity. He's the military philosopher of our hootch.

Well, you've heard from your number one son and the Winesap, so we'll close off for tonight.

July 30, 1968

Well, tomorrow another month's gone. I think maybe I'm a little calmer than I was the other night, and I'd like to soothe your fears about your Number One son. My wound was not all that bad. It was just two small shrapnel wounds, but the pieces were very small, maybe about one-eighth inch in diameter. They didn't do much damage. I've been walking around two days now without a limp. All in all, I was one lucky boy. It could have been lot worse. Had it been another inch, it could have taken my shin bone out. Of course, the HAC was the truly lucky one. If he had had his head turned the other way, it would have gone right through his face. Such are the fortunes of war.

I would like to make one comment, something rather humorous, and it is directed at you, Mom. In some of your tapes, you've been giving me little lectures on religion and stuff, and you made reference to the fact that I was clad in the protection of God. And I just have to say that on Sunday night, my leg must have been hanging out.

Anyway, I just wanted you to know I wasn't seriously hurt. It was kind of a minor thing to get a Purple Heart for, but I'm not turning it down. I'm gonna take it and wear it with pride. I am using it to my advantage, as I usually do. I have cut some slack for myself. I was immediately grounded because when I got back they didn't know what the story was, had to wait for the x-rays and everything. So I've been hanging around the sick bay for two days now, sleeping in late, and really enjoying myself. In fact, you ought to be rather proud of me. Two of us got together to see just how much slack we could cut ourselves from the system. For instance, the other guy, his name is J.S. High, an academy grad by the way, but a decent sort anyhow, was sort of the leader for this escapade. He and I were both assigned to do investigations. So we decided to just wait and see how long it took for them to notice we weren't doing anything. Well, our investigation was due on the 21st, and today is the 30th, and the report still isn't turned in. But the slack snapped today; the group called down and demanded to know where those reports were. So I guess I'll have to whip out a quickie tonight and turn it into them.

Our attitude is one of not caring very much. You may not appreciate that, Pop, but we're just not real pleased with our lot in life over here, particularly the way the war is being handled. I'll give you something to think about. Yesterday we got a call for an emergency MedEvac, only we didn't know it was an emergency because no one bothered to tell us. By the time we figured out this was not a routine mission, seven guys who started out as wounded in action became killed in action. That kind of thing gets to ya.

Another thing happened last night. We got a call for some emergency MedEvacs. Was gonna launch Bald Eagles. The section leader went out, got his bird turned up and was ready to go. But his wingman just piddled around and piddled some more until they were 40 minutes late. As a result, one of the wounded died in transit. The difference was probably that 40 minutes. What we are doing is pretty important work; I only wish all of my fellow officers appreciated the fact.

Last night we also had to extract a group of recon who had got into a rather hairy situation, and it was rather hairy to fly. People had to go into the mountains and land in the dark. It's hard enough getting into a lighted field at night. When you're trying to land in an unlighted zone with trees all around and guys trying to find you with a strobe light, well, it's a little hairy. But our guys went out, got the wounded and killed out all right. But let me tell you, those people were so thankful to us for pulling their asses out of the fire, well, it really makes you feel good about what you're doing. It kind of overshadows some other things around here that aren't so good.

In a way, I think we have some of the most dangerous work over here. I'm not taking anything away from the grunts. But when the bullets are coming, the grunts can hunker down and take cover. We can't. There's no place to hide up there. We just keep going up and getting shot at. It's weird, but that's our job. In fact, we don't get many awards for taking fire because it's so commonplace for a helicopter to take fire.

Well, today has been sort of slack for me. I slept in till about 10. Hobbled over to the coffee mess and counted my money, did a little work. Came back to the hootch and laid around in the air conditioning. I must say it was great to get your tape and hear all those familiar Barboursville voices. Even the ones who have trouble finding something to say, it's just great hearing their voices, I don't really care what they say.

You can tell my little brother I'm a full-fledged, bona fide, genuine combat veteran now, for what it's worth. And I'm sure when I get home, my Purple Heart and a dime will get me a cup of coffee. You know, the one thing about that whole incident that really bothers me, the thing I keep thinking about, was when that round came ripping through the cockpit. I can't really describe it to you. There are no words to describe it. It's worse than being in a car wreck. You know, I was in that one wreck, and you could see it coming. But with this, everything is all right and then all of a sudden, bap, this thing comes zipping through your cockpit. It's so sudden and violent, I think for about 40 seconds my mind didn't function. I just kept operating on instinct. It's all kind of blurry. Then gradually my mind started working again and we got out of it okay. And it's a matter of pride to me now that I did continue to function, that I can handle a situation like that.

August 1, 1968

Today I was designated pay officer. I had to get up at 5:30 a.m. and go to disbursement and pick up $26,000 and sign for it and then go down and pay the Squadron. Kind of a boring day. As usual, not everybody made it and so I didn't get to pay everyone. So I'll have to keep at it tomorrow until everyone is paid.

My life is pretty slow without flying and all. I hope to be flying again soon. I've answered some of the letters I've got, that's how bored I am. I have missed my Corvette. That beast is truly dear to me, and I'm sure my brother enjoys wheeling it around Barboursville. I've even dreamed about that beast since I've been over here.

We've gotten some new Majors in the Squadron and a new Captain, and it's kind of bad. Last night we got word that the XO, the dip shit, had decided the moustaches on his Lieutenants were kind of scraggly. Not the Majors' moustaches, mind you, only those on Lieutenants. So we've all been out waxing up our moustaches with whatever we can find. My moustache is a handlebar, and quite impressive if I do say so myself.

I have discovered I have a valuable position here at Marble Mountain in that I have access to beer sales. It seems that Marines are limited in the amount of beer they can buy. It would appear we have something of a reputation about that. However, as representative of the Squadron, I can go down to the PX and buy beer by the truck load. As a result, I find myself in an excellent bargaining position. I propose to bring home some souvenirs, maybe a rifle or two. It's amazing the loopholes one can find over here. And if it all serves to make life a little more interesting and take our minds off our troubles, well, so much the better.

Well, my roommate Bob W., just walked in. Bob, do you have anything to say to the folks back home?

"Hello folks. Just wanted to say hello so you can get a load of my voice. We all enjoy your tapes here. Just keep them coming. And we enjoy listening and watching Ace record them to send home. Hello and goodbye."

Well, that was Bob W. I can't think of anything else to say, so I'll close off for now.

August 3, 1968

My dear mother. I want you to know I truly appreciate your little religious messages you've been putting on the tape. But I think in the future, when you have thoughts along those lines, perhaps you'd do better to commit them to paper. First of all, it really takes up a lot of valuable tape space. In this last tape, for example, you said Dave and Pop would get to talk, and I was really looking forward to it. But they never made it.

Also, your messages have occasioned a good bit of hilarity here. The last time you mentioned that I was clad in a panoply of love, or whatever it was, and we all decided my leg must have been sticking out. You also mentioned something about angels watching over me. I looked up at that instant to see Tom W. searching the hootch. You guessed it, he was looking for angels.

Well, today I went down to sick bay. One of these pieces of shrapnel in my leg is giving me trouble. It's become a little infected. So the doctor said he would follow the old axiom that if it doesn't grow there, it doesn't belong there, and he tried to cut it out. But he was unable to find one piece, and I guess he's just gonna leave it in there. So after he bandaged me up, I went down to the Squadron and found I had been placed on the Bald Eagle, which is a night standby in case of a shit sandwich.

Well, I hadn't flown in about a week, and everybody's been giving me grief about it, so I didn't say anything about my leg. I'm just hoping we won't have to launch tonight.

Dave, I'm glad to hear you're having fun running around in my car. You might ask Sergeant H. to put you onto a place called Red's down in Huntington. We went down there when I was home after the bars closed and stayed until 5 or 6. Also, if you get a chance, let me know what you did in Thailand. We're pretty isolated here, and need that kind of information, since sometimes we get to take R&R there.

Well, I've gotta go down and pre-flight my bird in case I have to launch tonight. I sure hope I don't; my leg is hurting like a mother, but I don't want to duck my call. I'll check you later.

August 9, 1968

I got your letter today, Pop, and I thought I'd answer here on the tape. Sorry to hear my brother Dave's leave got cut short, but I guess he's realizing now he belongs to Uncle Sam. That's the way things go. I guess it was good for Mom to get to make that trip. I'm sure she'll enjoy it, and I'm equally sure it'll give you some peace and quiet back at the homestead. Something we can all use on occasion.

I was interested to hear the Barboursville social scene has picked up and they're having street dances. I'm a little curious to learn who was involved in these street dances. Teenagers I suppose, but it sounds like a good deal, sort of revs up the town spirit. You might be surprised to hear this, but I'd be glad to be back in Barboursville right now, no matter what they were doing.

I was glad to hear Paul T. made Captain in the Army. If you see his parents, convey my regards. Tell them I said that's good. I hope he's making out better than I am over here. I haven't seen him, of course, since he's way down in the southern part, and I'm up here in the north. But, we're both in the same boat, so to speak.

I would like to give you the rundown on my shrapnel wound. The whole problem with the thing is that they didn't try to cut them out the night it happened. I went back there and our flight surgeon was on R&R. One of the other flight surgeons looked at it and said: "Nah, just bandage it up and leave 'em in there. It won't hurt."

So a couple of days later one of them was starting to bother me when I was walking. So I went back and by that time my flight surgeon had returned from R&R. I told him that I'd sure like for him to cut them out, so he took a look and said one of the wounds had become infected. So he decided to go in and cut them out. The other one is a little larger, and it was the one giving me trouble. He was unable to find that one, though he could see it on the X-ray. But he was afraid to take a chance on doing damage to the muscle; said he'd rather just leave it there. Personally, I agree with you; I'd rather have it out. I kind of feel like if it didn't grow there, it doesn't belong there. But he said to leave it alone, for the time being anyway. Of course, it might prove to be a bonus if it enables me to get out of the country a little sooner, or maybe get some convalescent leave in Okinawa or Japan. I don't know. I'll just have to hang loose and see what happens.

The doc left the wounds open for about four days to let them drain. Then he sewed them up. They're healing slowly, still draining some, so it'll be about another week or so before we really see what's going to happen with it. In the meantime, I've been grounded. Been spending most of my days on the rack with my leg stretched out. Makes it feel a lot better. I've been enjoying the rest.

The rest of the guys, of course, have been out flying their gonads off. That kind of bothers me that I haven't been doing my part, but other than that it is a pretty good deal.

Well, tonight they said there was gonna be a show at the club, so I guess I'll have to make it to that. It'll be good for my morale, and anything good for my morale is welcome. Well, I'm running out of things to say, so I'll take a break now.

Later.

Well, it's late at night on the 9th, I think. The message I am about to put on the tape is one aimed entirely between me and the other part of my personality, the Ace. I would appreciate it if this particular part of the tape didn't play for anyone outside the family, because this is personal. I'm recording this because I never want to forget how I feel at this moment. I am completely and utterly shit-faced. And because of that, I am speaking the whole truth, that truth I will admit to myself only under these conditions.

Folks, the last time I talked to you in this condition was several days after a very good friend of mine was killed here. That is something I'm bitter about because that poor son of a bitch never had a chance. He didn't have his hands on the controls. His life was entrusted entirely to an aircraft commander that day, and there were a lot of people who had said it was only a matter of time until he bought the farm, and it was a tragic thing that poor Ray D. had to go down with him.

What I want to say tonight is basically this. Now I've been flying my part over here until about two weeks ago, but I went out one day and we literally got our asses shot off, and in the process I got hit. Now for about 18 days I haven't been flying; I've been sitting on my rack. And every one of those 18 days I've relived the moment that bullet crashed through my cockpit. Now I can't sit here and describe or try to make you see the picture of a bullet coming through the cockpit, because that's something I had never seen before, and something I hope to never see again, and it's something words cannot describe. But it was a terrifying experience. Now every day that I sit here and relive that moment, my asshole gets a little tighter, and it's got to the point my pucker string is about to break. The last thing I want to do right now is go out and strap myself in that airplane, because I know just as sure as I go out there again, those assholes are going to try to shoot me.

Now, they've shot at me before, sure. But they've hit me twice. I only took fire twice, and each time I took fire in the cockpit. Now, I ask you…who gives a shit whether I get wounded over here? Now, I know of course that my folks care, but I have to live with more than my family. I have to live with myself, and the world, my world at home. But I'll guarantee you one thing…my world at home has never been exposed to the things we are exposed to over here. If they did, it would affect their attitude about a lot of things.

And, I'll guarantee you one thing, when I come home, my patience with people who don't see the war in Vietnam the way I do, well, my patience will be extremely thin. I would not hesitate for a moment to whale into someone regardless of whether I stood to get my ass whipped. I would not hesitate for I have faced far worse things and come out on top. And if I come home in one piece, I'll never hesitate to step into something, because I assure you there is nothing at home to compare with what we face over here.

I remember a party I went to before I came over here. There was this one individual there who told me the only Vietnam heroes are the ones in jail because they wouldn't go. I should have grabbed that son of a bitch by the throat and just pounded his head in. Because he didn't know what he was talking about. He knew nothing. Only the people who have been here and have seen the waste and have felt the sorrow.

I've carried wounded to the hospital, I've carried dead to the morgue and I've seen more waste than I ever want to see in my life. I've seen enough here to feed nightmares for a lifetime and I don't need to see anymore. But for those narrow-minded people obsessed with their own opinions, and who don't care about the rest of the world, well the words I have for those people…well, I'd rather not use those words, but to put it mildly, they piss me off.

My friend is gone. It's so easy, so quick for someone to go over here, and there's nothing you can do. You just keep telling yourself that when your time comes, it comes. But it's hard to convince yourself of that. I tell you, I will be so glad to get back to the States, and I hope my little brother never has to come and risk his life in this screwed-up mess. I hope I'm the only one in my family who ever has to come to this place.

I got a letter from Jim the other day. He said something about this being a just cause, and that I surely must have felt patriotism, even though I joked about it. Well, I'll tell you something and I'll be perfectly honest. You really don't feel patriotism in this situation. That's one of the first things to go. That mother and apple pie bullshit just won't buy it. There's only one thing you're concerned with and that's the fact that your ass is in one piece and you'd like to keep it that way. That's your only motivating factor, that and your friends.

Now I have a family, a mother and a father, but I have no family of my own. The only thing I have in Vietnam are my friends, and right here and now my friends are the most important thing in my life. With them I live or die, and I tell you I couldn't have asked for a greater bunch of guys than the ones I'm serving with. Now, I don't know how you all feel about what's going on over here, but when I get home and have a chance to discuss it with you, I'm sure you'll see it the way we do.

Well, I don't know that I've really said anything tonight, but I don't really give a shit. The only thing I was really trying to do was put down on tape my feelings so that when I came back from this God-forsaken place, I'll never forget how I felt about it when I was here.

They say time heals all wounds, but this is one wound I don't want healed. I want it to stay fresh. Because if a lot more people had wounds like this, there's a better chance our society will be able to preserve itself. It's frightening sometimes to think that I'm wasting my time over here because eventually the communists are going to overrun our country anyway. But I'll guarantee you this, I'll never submit. I've already decided that if a law is passed that firearms are to be registered, I'll roast in hell before I register mine. I want you to take notice of that, the firearms at home are mine. I don't want anyone to register them, because I feel that's an invasion of my privacy. Also, if a police state were to take over, they could confiscate my weapons because they would know where to find them. But as long as no one knows where they are, I'll have 'em and I can fight back. If there's one thing I've learned from this experience, it's that I can fight with anything.

I live in a great country, and I'm proud of where I come from. I'd like to see it remain a great place to live, a great society if you will. But if it changes to the point that my individual freedoms are hampered, then I'll not hesitate to fight it. Just remember, folks, that if the law is passed to register firearms, you leave mine where they are, and you let me worry about it. I know there's a lot of things we don't see eye to eye on, and that we never will. I don't expect you to be with me on everything, even things like this that I feel strongly about.

I'd like to say one other thing. I came over here a bachelor, aged 25. Now I am a relatively old man in the conflict over here. I'm sure there's people that are older, but I'm talking in numbers. I've never had a chance to get a family started, and that's something I've wanted more than anything else. I only hope I'll be able to come back and find somebody and settle down and have a family of my own. I'm sure the experiences I've had over here will have been beneficial in some ways, and have helped me mature. I know you might not think so, but I don't have to face you in the mirror in the morning, I have to face myself.

Well, I think I'm gonna be sick tonight, but I feel as if I owed it to myself because every now and then a man has to get drunk. It has a tendency to cleanse your mind and body, and that's good.

Chapter 7
Back to Work

It is August 22 and it's been quite a while since I spoke to you. I thought you might like to know I've gone back to work. I started flying again about a week ago. The first was an admin run, and then a couple of days later I went out and flew with the skipper on a really bad day when the weather was almost below emergency minimums. We took in a team of four men to do some reconnaissance – two Lieutenants, a radio operator and a grunt. Not 15 minutes after we got them in there, they had some contacts, killed one NVA, and it looked for all the world like we were gonna have a problem on our hands going back in to pull them out. That would have been exciting, but as it happened we didn't have to do that.

The day before yesterday, I flew recon, the first time I'd done that in a long while. You may recall that recon are some of the hairiest missions we fly because the zones we go into are not secure. There are three basic things we do with recon – inserts, extractions and what we call flip-flops. That's where we take one team in and another team out. Actually, flip-flops aren't so bad because the team already on the ground secures the zone for you, or tries to.

Extractions can be bad, but most of the time they're not. The troops have been out there awhile, are cognizant of the area, and know what's going on. Of course, now and again the bad guys come chasing them, but most of the time extractions are fairly smooth.

It's the inserts that are the real bastards. You just never know on them. You fly over the zone, the Hueys go in, you run Fixed Wing prep, and you're still not sure. For instance, the day we flew recon we did one extraction and two flip-flops. Well, the first one went okay, but on the second, the Fixed Wing ran prep on the zone and got some secondary explosions. Well, to a helicopter pilot, that means two things. First, it means there is something down there that didn't grow there. Second, it means that whatever it is that didn't grow there was brought in by people. And the chances are that somebody is still waiting around hoping to bag a helicopter. So needless to say we decided not to go into that zone. All in all, an interesting day. Yesterday I was the Squadron boy, better known as the Squadron Duty Officer. That job is terrible. You sort of hang around the Squadron and go get people in the middle of the night. Last night, I had to wake up the stand-by crews because we got a call from group that we had to launch. A recon team had gotten into a bad situation and needed to be extracted while they were under fire. Fortunately, some other birds were available so our people didn't have to do it.

Well, well, it would appear my friend Bob has just come in with some bourbon. Bob has a smile of congenial hospitality on his face which means he will more than likely offer me some. And how could I not accept? Perhaps Bob would like to say a few words to the folks back home. Bob?

"Well, all I have to say, folks, is it was nice and warm today, it must have been a good 100 degrees, or maybe 101 or something like that. And when the wind blew, it brought along a little sandstorm. Enjoyed talking to you folks."

Well, that was Bob. Folks, I think I'm gonna take a breather while Bob uncorks that bottle. I'll talk to you in a few minutes.

Later.

Well, I'm back. It's been about an hour. Bob and I've been sitting here discussing what's going on in our Squadron. You'll have to excuse the sound of my voice. I've got a real bad head cold. My head's so stopped up I can hardly breathe. My sinuses are just packed, but I don't want to get grounded. I guess I'll fly tomorrow.

Anyway, a couple of things have happened here recently. As I told you, we've gotten some Majors here in the Squadron who are completely enthralled with the fact they are Majors, and we are but Lieutenants and Captains. One of the Majors decided that he wanted to institute a policy where co-pilots, when they reach the magic number of 500 hours, would automatically come up for the test for Helicopter Aircraft Commander. If they failed the test, they would go up before a Pilot's Disposition Board.

Now, the fallacy here is that there is no way you can arbitrarily legislate when a man is ready to accept responsibility for an aircraft and crew, and whatever he happens to be hauling. You can't sit down and say at 500 hours a man is automatically ready for it. Some guys are ready at less and some need 700 hours before they're ready. So I can tell you this, when I reach 500 hours, and if I don't think I'm ready, I will not hesitate to tell them I may get myself in a bind, but I will not take the risk of taking an airplane and a crew of four and whatever I happen to be hauling if I don't feel ready to do it.

Fortunately, almost every other HAC in the Squadron, with the exception of the Major, disagrees with this policy of the automatic 500 hours, so they have banded together to fight it. We may yet get it worked out okay; I hope so.

The other thing that happened – our skipper got grounded by the Group Commander because he went out to fly after visiting the club. So the skipper got pissed and said we were going to send 10 aircraft to the Bald Eagle every night. So this means from now on until it wears off or somebody gets smart again, we're going to have 10 airplanes and 10 crews on stand-by status every night. This means guys will fly all day and then be on alert status all night.

There's a lot of people up in arms about this, so I don't know what's gonna happen. It just seems like this chicken shit stuff never stops. We always hoped that when we got out of the training command, and got into an operational Squadron, that things would change some that we would be in the real Marine Corps. But it appears this is the real Marine Corps, which is not at all what I thought it was. It's always a bunch of Mickey Mouse stuff like this, and I am amazed that the Corps has been able to do as well as it has here in Vietnam. I really mean that.

Well, enough of that. I feel like I've been kind of lax on these tapes, and I want to get squared away. It seems though it's always just so much easier to go to bed and go to sleep but I've got a bunch of photographs and negatives to send to you, but somehow I never seem to get around to it. But I will try to do a little bit better in the future.

I enjoyed that tape I received yesterday. It was pretty good, although it sounded a little blurry in places, distorting people's voices. Mr. O. sounded a bit loaded, which doesn't seem in character. But all in all, I really enjoyed hearing everyone talk. I especially got a kick out of all those references to my accident. Actually, it wasn't an accident because it happened with malice aforethought. They were trying to shoot us down, so it doesn't really qualify as an accident, at least not as I understand the term.

Tell everyone I also relish the letters. It doesn't really matter what people say; I just love hearing from the home folks. I've been getting a few letters back into the mail. So tell everybody to keep 'em coming.

The last CARE package you sent was a bit hit. I got enough stuff there to share with my friends, except for the peanuts and cookies which I kept for myself. The mints went over well, and I made lots of deals trading them for other things. It was a real good box of goodies.

I would like to say it was nice of you to record the sound of my beast for me. I really look forward to getting home and driving it again. I will probably get some speeding tickets, not having driven in a long time. It's an interesting thing flying a helicopter, or any kind of aircraft, when you're not driving any kind of land vehicle. You tend to see motion differently. We're used to flying around 100 knots, sometimes pretty close to the ground. After a while 100 knots gets to seem pretty slow, so when you get in a car you tend to drive, not by the speedometer, but by the sense of relative motion you had before. But then that is one good thing about my Corvette – it will keep up with an airplane. At least it will keep up with a CH-46. Well, I can't think of much else to say tonight. So I'll close off and check in later.

August 23, 1968

Well, folks, it's five minutes after 2 a.m. and it looks like we have a little excitement here. There is right now a stiff firefight taking place about one and one-half miles to the south of Marble Mountain. There's a Special Forces camp down there. Flares are going up all around the area, a lot of aircraft are out, Hueys and Fixed Wing. They've got the ground troops out in force, and have sounded the siren for us to head to the bunkers. And here I sit in a ditch in the middle of the night in Southeast Asia wishing I were somewhere else.

Later.

Well, I tried to record some of the sounds from last night for you, but it didn't come through very well. Actually, there wasn't all that much to hear last night, except for the mortars. I thought I'd like to bring you up to date on what's happening around here today. I didn't really realize what was going on last night, but I found out today that within 300 meters of where we live, across the wire in one of the Army compounds, they captured three NVA in U.S. Army uniforms. South of Marble Mountain they had a lot of stuff going on last night, some pretty severe firefights. At one time, the NVA captured Marble Mountain itself and raised their flag. Some of our people went up there and got it, and now it's hanging in one of our hangars, a souvenir of the war.

About a mile and a half to the southwest of Marble Mountain, there is a river we always fly over at low level when we move out into that sector on business. But yesterday, the NVA took a couple of bridges down there which means we will now have to be careful flying along that river. All day there's been hard fighting over in that area. We were trying to take the bridges back, but I think we finally just decided to hell with it and dropped napalm all over everything. That's probably what we should have done in the beginning. I'll tell you, there's nothing any more beautiful than napalm; that stuff is fantastic. I just think Dow Chemical should make nothing but napalm. My roommate, Tom W., was out today and took two hits. He was telling me they were out flying around and around this village, like Indians circling the wagon trains, and his gunner saw some people running in front of a hootch. When they saw the helicopter, they ran inside the hootch. Well, the gunner just devastated that hootch with 50 caliber fire, one tracer, one armor piercer, and one incendiary in each sequence. And they figured they must have got everybody, but Tom said pretty soon they saw people running around again.

Down south we could see the Hueys using the new mini-guns. I don't know if you've heard about them, but they have a tremendous rate of fire, something like 6,000 rounds per minute. It doesn't sound like a gun at all, more like a buzzing sound. So all in all, this has been an interesting day around here.

I didn't have to fly any combat missions today. We did make an admin run down to Phu Li, and about 40 minutes after we left they started taking fire in there, so I guess we got out just in time. Tonight should be interesting. They're expecting quite a bit of action and have warned everybody to keep their side arms handy, as well as our flashlights and helmets, and take them with us into the bunkers. A very prudent suggestion.

Tonight I'm on the Bald Eagle stand-by again. I'll be in the first plane of the second section, so I probably won't have to launch. I hope not because it would be a guaranteed shit sandwich. In fact, the birds that were flying early this morning were having to fire into Marble Mountain from 2,000 feet, something I've never seen before, and a lot of people who've been here longer than I have said they've never seen anything like this at Marble Mountain. It's always been considered a pretty secure area. Anyway, there's a bunch of NVA in the area, so it should be an exciting evening.

That evening…

Well, folks, it's night. I'm sitting here in our bunker with about 20 other people watching the Spookies work right off Marble Mountain. We have an enforced blackout this time. All the power's been turned off so no one's hootch has any lights on. I had to go down to Squadron, thought we might have to launch, but we got out of it. We did launch the first bird, but I was in the second and didn't have to go. But we're still on one-hour stand-by, and apparently the whole world has turned to shit in the Marble Mountain area. I don't know if you can hear those explosions in the distance, but they're quite audible to us. Now someone just said there were 4,000 g**ks around the perimeter, that we might get overrun tonight. That noise right there is a Spooky working nearby. I've just been told the reason we don't have electricity is because one of the canister fires hit a transformer and destroyed it.

Well, that's about all for tonight, or at least I hope so.

August 24, 1968

Well, it's been pretty exciting around here, everyone is out killing people. They've been using artillery and air strikes. I saw two F-4s dropping 500 pound bombs on people. I listened on the radio while two Huey gunships rocketed some people, and then when some of them started to run away, he rolled in and machine gunned them. So as you can readily tell, it has been a most productive day in the An Hoa Valley.

I went over to Hill 55 today, into their briefing room and I was sitting there looking at the scoreboard up on the wall which lists total NVAs and VCs killed and wounded. And then they brought in this chart that lists all the units in the field, and showed the kills they've got against the NVA and VC, and the only thing I could think of was that it was all a big game, and these were the An Hoa Valley Conference scores.

The only thing we had to do today was to take two different groups, new officers, on an overflight of the area so that we could show them the boundaries. They need to know the boundaries because if you step over them, you get penalized. They take a day off your R&R, what a deal!

Well, you've probably heard on the news that the war around Da Nang has been hot and heavy of late. They were expecting Marble Mountain might get overrun, but as of yet the NVA has been unable to muster sufficient forces to break through. I think if we could just get them out in the open enough we could do them in. There's no question we have the firepower to kill those people in droves. The only good VC is a dead VC, and the same applies to the NVA.

Tom W. came in this evening. He's been out flying around. He saw an artillery duel between the NVA and local Marine units. Of course, it's hard to tell at night who's winning. He's also taken heavy 50 cal fire at 4,000 feet, and he also said they saw hits taken from anti-aircraft guns. We've been trying to take them out, thus far with no luck. Things like that don't help your morale much.

Tomorrow I've got the day off. I won't be flying, not for a while anyway, until they discover they need someone out there. Then they'll come and drag me out of bed and say, hey, you've gotta fly. And I'll be pissed. But until that time, I can rest easy and enjoy myself, something that doesn't happen every day at Marble Mountain. That's about all for tonight. Catch you later.

August 25, 1968

Tonight I'm sitting here with my roommates over a continental style meal. Bob W.'s wife sent him a lot of delicacies wrapped up in cans – some cheese, some Japanese shrimp, crackers, and stuff. We're washing it down with beer and reminiscing of earlier days. We're eating in the hootch instead of the mess just to do something different. The mess has the same thing every day to the point you can tell what day it is by what's on the menu. Also, the quality of the food in the mess has been going down of late.

Well, today's been pretty slow; I didn't have to fly. I laid around in my rack most the day looking at magazines. A pal came by and we discussed women, cars and motorcycles. I've decided to sell my motorcycle when I get back. You know why? So I can buy a new one.

Just kidding. I'm not going to sell my motorcycle. I just thought I'd tell you that and get you all worked up. Also, I'm going to keep the car. I miss them both terribly, and can hardly wait to get back to crank them up.

Our hootch maids came back to work today, so the war must be easing up. You can always tell. When the hootch maids don't come to work, it means the VC are lurking nearby. Our hootch maid just told us she saw the ARVN forces kill three Viet Cong. I'll tell you what; we've decided we just don't like this place at all. Coming home is just like something that is never going to happen. It is going to be such a great day when we do get home. I may be several days getting home after I reach the west coast because you know in airports I tend to socialize a bit, and it's easy to lose track of time and miss your flight. I don't know if I told you I missed three flights out of Chicago when I was on my way over here because I was talking to this Army helicopter pilot in an airport bar. So considering the state of mind I'm in, it might take me a long time to make it from California to West Virginia. But fear not, for I shall make it eventually.

I might add – one way or the other, but that's not nice. Tonight I'm on the ninth section of the Bald Eagle. All that means is I have to stay in the hootch and can't go to the club, so that's not so bad. For all practical purposes, I just have to stay in my hootch.

I maybe could get a couple of letters off tonight, but I don't know. Somehow letters are getting real hard to write. I just don't know what to say to people. Of course, that's not exactly a new problem for me. Sometimes, though, I get real prolific and dash off a whole bunch of letters at once. Well, I'm just babbling now, so I'll sign off for tonight.

August 27, 1968

Well, folks, I'm here to tell you there's never a dull day when you're flying with the old magnet Astle. It's only 3 in the afternoon and I've already had a full day. Got up this morning about 4:30, supposedly to fly recon. However, first we had to go out and recover the bodies of a Huey crew, a Huey shot down last night. They sent some people out at the time and managed to get the pilot, but not the rest of the crew. The pilot was badly burned, but is expected to live. The others were burned beyond recognition.

We left here before daylight, got out to the area, picked up our two Huey escorts, and went out. The Hueys didn't know exactly where the flight had gone down so they had to search the area. We had 15 men on board as did our wingman, and we had Fixed Wing on call. We were supposed to go in, recover the bodies and any other equipment that was salvageable. Well, we finally found the wreckage, the Hueys got it all marked for us and everything, and the first bird went in, dropped off his 15 men, and lifted out. And on starting out he took a bullet in the cockpit just behind the pilot's head.

Well, we climbed a ways and talked it over. We decided there must be a sniper down there, and that we'd go ahead and drop off our team. We started down and when we got to about 1,000 feet we started hearing a popping sound, the same as when you're popping popcorn, just a horrendous sound. I had never heard ground fire before. I had heard some guys talk about hearing it, but I never had, even though I've had an aircraft hit several times. But today I heard it plainly.

Anyway we continued down, got to about 300 feet, turned 270 degrees to the east and headed for the drop off point. And just as we began to slow up, we started taking hits on the aircraft. Well, shit, we put on every conceivable bit of power we had to turn and avoid the fire, but we were taking fire from all sides and there was nowhere to turn to get away from it. So we just bored right straight on out at low level straight back to the secured zone about 3,000 meters away. Well, we got into the secured zone and shut down to discover that three of the 15 troops had been wounded, one of them badly. And the damage we had taken was just amazing. It was no doubt the worst I've been shot up over here, and my pilot said it was the worst he had been hit, and he's been flying over here for 12 months. We counted at least 12 holes in the aircraft; I don't know if the crew chief found any more. But every time we went over the bird we would find more holes we had missed the first time around. They were indiscriminately scattered all over the aircraft from front to rear, except for the cockpit, which was all right with me. We took one hit in the main spar in one of the aft rotor blades, so we had to just stay there until we could have some new blades flown out to us. Ruining a spar just destroys the effective stroke of the blade, and it's very likely to break off, which is not a thing you want to happen. We took another hit in a forward blade that wasn't in a spar, but it was in one of the ribs or something. I don't know exactly what they call it, but we had to get that blade changed.

We had another blade on the forward head that had taken a round through one of the pockets. That in itself doesn't make any difference; you can still fly it. With a hole in the pocket, it just makes a whistling sound. Then we took a burst of three rounds through the upward forward pylon. They were about three inches in diameter. The guy must have a rapid rate of fire to put that many hits that close together. We had one from up on the left hand side about midway in the cabin section. It came through the bottom and imbedded in the overhead. We had one round come through the side that was the one that seriously wounded the guy. And, well, you get the picture. Bullets can do a lot of damage to helicopters and humans.

It was rather exciting. Can't say as I enjoyed it a whole lot. But it was interesting and I wasn't frightened this time. Actually, it all happened very fast, probably no more than 45 seconds all told. I didn't know at first we were taking fire, even though I heard it, because I didn't know the sound. It wasn't until I heard the rounds hitting the aircraft that I knew we were in serious trouble. Fortunately, though, the bad guys were a bit premature in their firing. If they'd waited another 30 seconds until we'd slowed down and were actually in our approach, they could have just nailed us to the wall. But as it was, we still had about 80 knots, although we were about 300 feet, and were able to punch over the nose, pull maximum power and get out of there quick. So we were pretty lucky, although some of the guys in the back weren't so lucky.

Later on they called in three Phantom jets, F-4s and they did just a beautiful job of blasting the hell out of the area. They were dropping 500 pound bombs, dropping napalm and firing rockets and 20 mm cannons. They just had a daisy chain. In other words, the aircraft were flying in a circle, one rolling in right after another. As soon as one made his pass, another would come in. Just one continuous thing until they had all expended their ordinance.

Then the bird we were flying wing with, went back and extracted the 15 men he'd dropped, along with the bodies of the crew from the Huey. So the mission was accomplished. Our leader went ahead back to Marble, picked up another wingman, and went on to do the original recon mission we had been slated to do in the morning before we were sidetracked. My pilot and I came back to Marble Mountain and secured for the day. We decided we had earned a rest. I don't like that stuff at all, but it wasn't the last time I'll be shot at. It wasn't the first time and it won't be the last.

The guys are starting to kid me now like I'm a walking impact area. Some of the HACs are saying they don't want to fly with me anymore because I attract bullets like a magnet. I don't know about that, though I think I do have the high for hits on a co-pilot. That's a dubious distinction, but a distinction nevertheless. That's about all for now, but that should be enough.

September 3, 1968

Jeez, I'm getting almost as bad at making these tapes as I am at writing letters. I might try to get on the stick here a little bit and get some more done. Not a whole lot has happened since I last spoke to you. I've been doing a lot of paperwork, bureaucratic stuff. I flew yesterday for the first time since the 27th. We helped another Squadron off-load from a ship. It was something different; got to make a couple of landings on a carrier. Actually, it's not a carrier like you think of a carrier. It's what they call a LPH, or Landing Platform Helicopter. It's designed to carry a battalion of Marines and all their gear, plus a Squadron and its gear and airplanes. I guess we'll be hitting the boat, or going out to the boat, probably sometime in latter December, the way it looks now.

Well, it looks like I'll be going on my R&R in January. They sent out the word there will be only one R&R per tour. I put in for Sydney and it shouldn't be any problem. I should get it and that should be a good trip. I got the photos you sent today; thanks very much. I would like to place an order for one of those photo albums. I've taken quite a few photos and am running out of album space. I got the package you sent, and would like to make a few comments. The dart board has gone over real well. It's the most popular item that came in the box, with the exception of the moustache wax. We've played with the darts a lot, though one night one of the pilots came over here drunk and broke one of them, so now we only have five instead of six. So next time maybe you could send a set with the metal base that might hold up a little better in hard use.

Last night we spent more than an hour throwing darts, and the guys also got a kick out of the comic books. They come in here and read them aloud with feeling; it's hysterical. But I don't think the coloring books will get a lot of use. They are humorous, but coloring is not a big thing with Marine Officers, leastways not in your typical combat situation.

And oh yes, the pucker strings were okay. One never knows when a backup pucker string will come in handy.

Chapter 8
193 Cases of Beer

It is the fifth of September and I thought to add a few lines to what I recorded for you a couple of days ago. I perhaps did not mention that on the same day I received the package from you, I got another one from a, uh, female friend of mine, that contained a bunch of Hickory Farms food. It was pretty good. In fact, one of my roommates stayed with me for dinner in the hootch last night instead of going to the club, and we really enjoyed it.

Today, I flew a recon and it went well; didn't even take any fire. It was a lot of work, and there were some scary parts, but we didn't have any special problems. And then this afternoon I served on a court martial, one of six officers on it. Then one of us was challenged and had to go, so then there were five of us left. It was a tough case. A Marine charged with being in an Army club and threatening one of the bartenders, and also stealing some beer. And then the manager caught up with him and accused him of stealing the beer, and he allegedly assaulted the manager. Also, he was accused of disobeying a direct order in that Marines here are specifically prohibited from entering Army clubs.

The prosecution had two witnesses to prove the guy had been in the club and stolen the beer and done all the other stuff. Then the defense came out with six or seven witnesses who claimed the buy was actually in the Marine club at the time; they had a floor show that night and this guy was allegedly there drinking with his pals. So what do you do? Well, I didn't feel as though I could vote for a guilty verdict seeing as how there was all that conflicting testimony. The defense witnesses swore the guy could not possibly have been at the Army club doing all that stuff at the time the other witnesses said he was. So the guy got off scot-free and was completely exonerated of everything.

Yesterday, I served all morning on a flight audit board, the enlisted flight audit board, which is a job that requires you to audit all the enlisted flight times to make sure the master log, the aircraft yellow sheets and the monthly sheets all tally, and that all entries have been made so nobody gets screwed out of flight time. A lot of times the clerks forget to log a hop or something and get it screwed up, so we have to keep close track of all the action.

The rest of the afternoon I spent at my coffee mess trying to get my books balanced. Boy, what a struggle that was! But through perseverance I finally figured out what I was doing and believe it or not I finally got them in balance. So maybe I'm not doing so badly after all. It rained all yesterday and all night and is raining still today. A typhoon is hitting just a little north of Da Nang, and we're getting some effects of it. We had winds gusting up to 50 knots and it's been raining like hell. I've never seen anything like this back home. The wind is blowing wet sand up about head high. You get sand in your eyes, wet sand. And of course no one went flying today; we didn't launch any aircraft. However, some grunts got themselves in a bad situation because they don't plan for what's going to happen. They had 235 troops and four Howitzers on this little river island which flows out to the China Sea just a little bit south of here, and they were afraid they might drown. So one of the other Squadrons launched four 46s, and then one 53 and two gunboats, to bring them back in. So it must have been interesting trying to fly in this weather. I watched a Huey this morning that was flying almost 90 degrees off his track, or his course heading, to maintain his course because the wind was so strong. And he was only making about 20 knots of ground speed, getting knocked all over the sky.

Our next door neighbors, the guys who live in the other end of our Quonset hut, have an inch of water on their floor. We've been more fortunate; the roof on this side doesn't leak so badly. We only have one small leak and thus far have handled it with a wastebasket. I really feel sorry for the grunts that have to go out in this downpour. In particular, I do not envy the guys out there manning the watch towers all day. We had to tie down the latrines to keep them from blowing away, and of course we had to lash and sandbag the aircraft. Some of the older hootches had to have sandbags on their roof to keep them from blowing off. So we've all been pretty busy, and we are all wet.

I got your last tape today and would like to comment on a couple of things. First, we have plenty of music available here, a good radio station in Da Nang. The guys said you must have recorded that music because it sounds like World War II stuff. We are more inclined toward the wangitty-wang type music. I appreciate your efforts and all, but in the future that's a negative on the music. Also, the comedy routine you recorded was blurry and hard to understand. So negative also on recordings off television. Now, some of the guys here have TVs, and we get that Rowan and Martin's Laugh-In off Armed Services Radio Television Network, so we can watch that on the TV with no problem.

Other than that, I really enjoyed the tape, hearing your voices and all. But then I must insert another negative on the lectures. Of course, you know how much I enjoyed those when I was home. And I really hate to have to put the recorder on fast forward for fear I'll scoot right on past the end of the lecture and miss something I wanted to hear.

I've been told I will make HAC, or Aircraft Commander, on October 15, which gives me about 40 days. To tell you the truth, I'm scared of it. This week I flew some recon with one of the outstanding pilots in the Squadron, and it gave me an opportunity to shoot several approaches, one of them into a recon zone where we were doing a flip-flop. That's when we take one team in and another one out. That's when they want to keep a team in an area continuously. It gives the guys a chance to come in every four or five days and get some rest, because it's pretty hairy work they're doing. Then I shot another approach into another zone. I was a bit worried because they had been taking mortar fire when the aircraft came in. And in truth, if I had not been flying a Delta model which is more powerful than a 46 Alpha Model, and if it had been a couple of degrees hotter, I'd have probably crashed it. I really screwed up my approach pretty bad. But I got to thinking about that, and when I make HAC, there's not going to be anybody around to pull me out, so the first 25-30 hours are going to be pretty hairy. And I imagine I'll have some stories about that to curl your hair.

But if I make it through that first 25-30 hours, I'll start getting the feel of it. One bad thing about flying as a co-pilot is that you don't get to do that much flying, and you don't really know what goes into an approach. Over here when you get over a zone, you get a zone brief, and the grunts on the ground will tell you what the enemy situation is, where you can expect fire to come from, and they'll suggest a direction of approach. You discuss this with your gunbirds. You decide what you're gonna do and tell them so they can cover you when you're down there in the zone. But you have to take into consideration the wind direction, where the enemy positions are, how you're going to spiral down, what direction to go in from, and also what route to lift out. Then when you get started you've got to play the aircraft against the winds, the rate of descent and the air speed.

In the two approaches I messed up, I had the winds figured out pretty well, and I figured out my route in, but I misjudged my rate of descent and my air speed. As a result, when I got down closer to the zone I didn't see that I needed corrections until it was a little bit late. I came into the zone high and fast and when you get into a situation like that and try to save it, you really get into deep trouble. But I think that after I've had opportunity to fly approaches for a while, I'll be able to hack it okay. I will say this...since I've been shot at so much over here, I like to go down below where we normally fly, which is about 3,000 feet. That is above small arms fire and keeps you pretty safe from 50 calibers. But I like to get from 3,000 down to the deck just as fast as I can. In other words, I keep my air speed up to about 110 knots and I like to get my rate of descent up to about 2,500 feet per minute, or more if I can get it.

But what I haven't learned to do yet is to judge at the bottom when I've got to start slowing my rate of descent and start slowing my air speed. I have a tendency to keep barreling in there to the last minute. This is something I need to get over. I think once I conquer that fear and learn to just fly the airplane, and forget about being shot at, I will do a lot better. So as I say, that should be pretty interesting when the time comes, and I'll have some stories to tell, I'm sure.

Now Pop, as for maintenance of the Corvette, just go by the driver's manual. I really don't remember much about it, but all I ever did was follow the book. As for a wheel being out of balance, I don't know. You say it wobbles at about 60 and I really wouldn't know. I rarely ever ran it at that speed. Whenever I was in that range I was just passing through. But I guess it could be. Go ahead and have it checked. I'm not worried about it. I'm sure in fact that you are taking better care of the car than I ever did.

I can only say I am really looking forward to climbing into that baby again and pushing the accelerator to the floor. I sometimes wish these helicopters had the same getaway that beast of mine has. I'd feel a lot better sometimes. You get in a bad situation; you can get out a lot quicker. Just burn rubber and make it.

Oh yeah, if you should run into Tim G., I have a message. You may not recall, but we saw him just before I left Barboursville, and he said the war was almost over, that we had them whipped and on the run, that they we're just using young kids to fight now. Well, I'd like for someone to put Tim straight and tell him that's so far from the truth I can't believe it. These guys are getting meaner than snakes over here. This thing looks like it could drag on forever.

It is my considered opinion that the NVA are capable soldiers, but they make a lot of tactical mistakes. We could do them all in, I believe, and end this war in no time if it weren't for the restrictions imposed on our troops. For instance, suppose I'm flying low and start taking ground fire in an area that is not designated as open. I can't let my gunners shoot back. I have to call Da Nang and they have to get in touch with the village chief, through several layers of command, all the way down and then all the way back up. Hell, by that time we are either dead or gone. Also, the bad guys will not wait around for all this to happen.

The 50 caliber machine guns mounted on our birds are pretty effective, and they will discourage someone on the ground from shooting at you, if you're in a position to fire back. Of course, we try not to fire a lot in areas where our own troops are around. We don't want to shoot up the friendlies. But if they would only take the restrictions off our troops, I'm convinced we could nail them to the wall. No problems. Sure, we'd take some casualties doing it. But we're taking casualties now, and it would be better than dragging this thing out for seven or eight more years, which is apparently what we are doing.

I heard today it will be several more days before I get any mail; the weather's so bad that mail flights can't get in. By the way, in your next CARE package, please send along some cartridges for my pen. I ran out and need some more. They don't sell them here, only refills for ballpoints and I don't like ballpoint pens.

I've put in for R&R in January. A lot of guys are taking off already because they are so fed up, but I figure it I'll be even more fed up by January, which will be about half way through my tour. I'm going to Sydney. I guess I could go to Hong Kong or Bangkok, but to tell you the truth I want to be among some round eyes who speak my language and are not shooting at me. I think it will be a lot more relaxing and enjoyable.

You can tell my brother, when I get back he and I will go up to Salt Rock with that boat Aunt Sue gave me and put it in the river, just spend the whole day floating down the river. That would be real peaceful and quiet. Although it may sound like a paradox, I also want to take along some guns and shoot up anything that makes a likely target. Take out some aggressions, maybe. Tell him to be looking forward to it, because I am. Maybe we'll take Jim T. along with us.

Well, I'm out of things to say. I'll talk again when I get the chance. Adios, amigos.

September 9, 1968

Well, not much has happened in the last week. In fact, today's only the third day I've flown this month. Three days out of nine isn't so good.

Last night we had a real fine show over at the club. A Filipino band with two round-eyes in the show, and a round-eyed MC who also sang. Truly and outstanding show. Also, there was a large group of Army nurses on hand and after the show we had a big party in one of the hootches. It was good to talk to round-eyed women again.

This afternoon I went flying. I was on a stand-by crew for the whole day and finally had to launch this afternoon. Flew up to Phu Bai and dropped off some passengers. Then we went up and flew over Huế and the citadel where they had the big battles during the Tet Offensive. Kind of interesting place to see from the air. From there we flew up to Quảng Tri and then to Đông Hà, which is in sight of the DMZ. We saw several landmarks up there you may recall from the news, such as Hill 881, what they call the Rockpile. There was a tremendous battle there.

Anyway we were carrying externals from Đông Hà to a place called Mỹ Lộc, which is an ARVN compound. Then we had a leisurely flight back to Marble Mountain. I got about 4.1 hours flight time. And when we got back to the ready room I was surprised to see the mail had come. I guess I made as good a haul as anyone in the Squadron: a package from you all, a package from Grandma, a package from Aunt Sue and the envelope you sent me. Also got letters from Grandma, Pop, Davie S., Larry, and a card from Anna Fay. And I got the tape you sent back, so I made quite a haul; really lifts your morale. Couldn't ask for better.

Now I must tell you about the latest letter I got from Miss B. Now, I don't make this a common practice, but this letter I just had to share with the troops. I invite over my friend Steve H. You know him; he's the one who took such a keen interest in your comment about how we're protected by a panoply of love. Steve said he preferred a panoply of steel. Anyhow, Steve just loves these letters, mainly because they have no punctuation and the spelling is so atrocious. Also, some of the personal comments are rather humorous in light of the fact I only met her a week before I left. Anyhow, Steve sits here and reads them aloud just as they are written, and it just puts the guys rolling on the floor. I'm not kidding; it's absolutely beautiful. So I've got to encourage her to write some more. Her letters are just priceless.

"Mrs. Mother Ace, your chocolate chip cookies are great, as you can probably tell because I'm chewing one right now. Really good, and thanks a lot."

Well, that was my roommate Tom, the one who prefers the panoply of steel. He considers himself something of an expert on chocolate chip cookies. He just took another bite, has a smirk on his face and now there he goes to the floor, clutching his stomach. But it was good while it lasted, and you can keep those chocolate chip cookies coming.

Last night my other roommate, Bob W., decided to get loaded during the show, the first time he's been drunk since he got to Vietnam, which must be some kind of record. Anyway, he really did a job on himself and got sick. So then of course they came in at zero dark 30 and told him he was flying this morning. He was really hurting. In fact, they had to stop the airplane three or four times so Bob could lean out and barf. They would just hover there at 5,000 feet so he could barf without getting it all over the plane. But he had a real good time which is the important thing. Occasionally you need something like that just to let off steam. I don't know if you realize it, but there's an awful lot of pressure in the work we do over here, particularly in flight crews. As long as you're flying, the pressure is really on you. And you don't have many opportunities to let off steam.

Well, I'm out of stuff to talk about, so I'll take a break.

September 12, 1968

I got a tape from you today, and frankly I am appalled at the waste of good tape space. The entire back of the tape was a television program. The whole hootch is appalled! The whole world! The whole Marine Corps! I told you on my last tape that we really don't need radio or TV programs because we get that stuff here. So from now on that's a negative on radio or TV programs, or any other kind of recording. I want to hear voices or my car engine. You can have Pop rev up the corvette for me now and then. So in the future keep it on necessities and we'll all be a lot happier.

Well, we just got back from the show and it was pretty good. They had three Korean girls, and I'm telling you, two of them were simply beautiful. I couldn't believe it. It makes me wonder how I'm gonna make it for six months until I get my R&R in Sydney, between now and then without female companionship. We've got our hootch mates here, of course, but somehow they don't fill the bill.

We flew this morning on a strike mission. That's where we put troops in where we think there's some NVA or VC to get on with the business at hand. Well, today we took in 150 ROK Marines, that's Republic of Korea Marines, and dumped them in there. It was pretty exciting. At first I was a little bit scared, to be honest, cause we didn't know what would happen in the zone. But as it turned out, we didn't take any fire and it was lots of fun. Of course, even though you don't take fire on your first couple of approaches, you still fly the succeeding approaches as though you are going to get fire because you just never know when you will.

After we completed that, our gunbirds left and we were supposed to go back and to re-supply for the ROK Marine outposts around the area. Now in the past we have some problems with ROK re-supply because we tend to get shot at a lot. And it's sort of a hard and fast rule around here that you don't fly into any of those zones without Huey gunbirds. We'll have anywhere from two to four 46s working with two gunbirds. The gun bird is a Huey with four machine guns mounted on it. Two are mounted in front and are fired by the pilot and two are on the sides.

Beg your pardon…they have six guns. Two on the front operated by the pilot, two on the sides fired by gunners and two more mounted with the rocket pods on either side which I suppose are fired by the pilot, or maybe the co-pilot. So you have six machine guns plus rockets…a lot of firepower. Now they've developed their tactics to support us in pairs, and they're quite effective. They can follow us into a zone, give us covering fire and cover us when we lift out. It requires coordination, but it's not all that difficult. It's actually pretty easy if you use your radios and state your intentions clearly. And we're the boss; we tell the Hueys what we're gonna do and they give us cover. But if you cooperate with them, you get outstanding coverage. It reduces the incidence of 46s getting shot up when we use the gunbirds properly.

Anyway, our gunbirds went home because they had been fragged for another mission. So we went back, refueled and returned to the pickup zone, which was secured. I was flying on the wing of the Squadron CO. He decided to get on with the job even with no gunbirds there because the zones were secure and all. So we went on and nobody got shot, which is good.

The guy I was flying with is one of these new Majors who was a jet pilot but he's caught on to helicopters pretty well. He was real enjoyable to fly with because he split up the hop evenly with me and I got to shoot about as many approaches as he did. It helped because I'm due to make HAC pretty soon and I need the practice.

If you see the Scotts before they leave, tell them I'm going to Sydney and I'd like to see them when I'm in Australia. If they'd like to come down to Sydney, I'd be happy to spend some time with them during the day. Sydney's supposed to be a real nice town, and I intend to make the most of it.

Well, I had another interesting thing happen today. It was the second day in a row I've been in trouble with the Executive Officer. The day before yesterday, he called me into his office and made me submit a report on a party that the Squadron had back in August. Since I was Special Services Officer for the Squadron, I had to get everything organized. I had procured 93 cases of beer. Well, actually it was 101 cases, but we got eight back after the party was over. And he jumped all over me, said he didn't see how it was humanly possible for 200 people, well, he said 180 people, to drink 93 cases of beer. He wanted to know what happened to all that beer. Well, I told him the only thing I could think of was that they drank it. But that didn't satisfy him. He said: "I WANT YOU TO FIND OUT." Yessir, yessir, three bags full.

So I went out and talked to some of the staff NCOs that had been at the party and told them the problem, so we got it worked out. And this is what I told him when I went in there today. I said, well sir, 93 cases of beer is exactly

2,232 cans. I checked with the Sergeant Major and discovered there are approximately 240 officers and men in the Squadron. You divide 240 into 2,232 and it works out to about 9.5 cans per man. Then you figure that a lot of them went to parties both days, and you take into consideration that a lot of guests came to the parties, quite a few in fact, and also the clean-up crews found a lot of half-empty cans lying around. So that kind of stopped him for a second. Then he looked at me and asked, "HOW MANY CASES OF BEER DOES THE SQUADRON HAVE IN THE COFFEE MESS?" And I said eight cases. And he looks at me and says, "HOW MUCH OTHER BEER IS THERE IN THE SQUADRON?"

Now a little background here before I go on. I've been taking orders for beer through the coffee mess. The officers and men come to me with their orders. I take their money and when I go on my supply runs, I pick up their beer for them. Then I keep it in the coffee mess until they come to pick it up. It's just a convenience thing. In fact, the CO had initiated it way back when. But now the Executive Officer decides he doesn't want beer in the Squadron, God only knows why. He's only got eight days left; I don't know why he would wait until now to start raising this stink.

But anyway, he asked me how many cans of beer belonging to other people were in the coffee mess and I had to tell him I didn't know because I'd just got an order in. So he jumps up from his desk and starts yelling and screaming and the next thing I know he's giving me five minutes to get it all out of there. So I say - yessir, yessir, three bags full, and went running out of there. I found out later he followed me to see what I did. Anyway, I went over to the mess and couldn't believe my eyes. I had 140 cases of beer to get out of the coffee mess. So that took some real doing, but I managed to get it all out. And I haven't been back to see him. I figure if he wants me, he can find me.

The guy who's going to be the new CO is a lot easier to get along with. I think we'll probably go back to the old policy, and I'll be Beer Procurement Officer. I never realized how important beer was to fighting a war until I got over here. I think they should put women on the same status with beer, but that's another subject. Well, let's see what else I can think of to report. I will say this, I'm becoming a little better adjusted to this place. In fact, today I was really enjoying it out there. I had a good time and hated to give the stick up. I just wanted to fly it by myself for the whole hop. We were out for five hours. Five hours in the cockpit is probably equivalent to 16 hours on any other job, at least. Wouldn't you say so, Tom?

"I would say five hours in the cockpit is equal to about 16 hours in a civilian job."

I thought you'd probably say that. I was going to come home from the club and write a letter to some honey tonight, but…wait a minute, now Bob has something to say.

"You can't compare different jobs. It takes a certain kind of guy to sit behind a desk, and a certain kind to sit in a cockpit. If we had any brains, we wouldn't be here in this God-forsaken place. Let's face it, Vietnam, well, I won't say it because Mother Ace is listening. But if we had our druthers, well, let's face it, just look at this situation in Vietnam. Okay, we're fighting against the people from…….ahhhh, who gives a shit?"

Okay, Tom's writing a letter to his wife, so we're just going to let him finish it. Anyway, you know how it is. When we get home huh…wait now. Tom has something to say after all.

"The problem when you discuss this situation with intelligent people like Ace and Bob is that if you're optimistic in your viewpoint, and if you're objective, your views are constantly changing. You run into different situations every day and after a certain amount of time, your views mesh into an opinion. We've only been here three months, and I know the other guys' opinions have been changing, and so have mine. I've run into some good people…Vietnamese people…that deserve our support. I've also run into some real assholes over here. And I say to myself, is it worth it for me to risk my behind for some worthless farmer who could give two shits what happens to his country, or to my country. It all depends on your outlook. One day you might be talking to somebody that you think is definitely worth your efforts, and another day you run into someone you think is a complete asshole. Of course, you also run into assholes in America. But all things considered, I'd rather risk my behind for somebody in America."

Well, so much for Tom's views on the war. Of course, it's something we discuss frequently. In fact, it's foremost on our minds since we spend much of our attention on the war effort. And we're not real pleased with it, as Tom just told you. Anyway, as I said, I was gonna come back to the hootch and write a letter to one of my honeys, but I just couldn't think of one who was worth the effort. In fact, I've thought a lot about that. And I just can't think of any right offhand who are worth the effort. Sure would be nice to know one though, it really would. Well, I think I'm gonna grab a shower and go to bed.

Chapter 9

These Worthless People

Today is Monday, September 16, here in good old RVN. Yesterday I did the most thrilling thing one in my profession can do, I flew a strike mission. This is something completely apart from the rest of our work. It was exciting and a tremendous amount of fun. A strike is when you lift a bunch of combat troops into a zone. Used to be we ran 'em up on shore in landing craft; these days we set them down with helicopters.

We had about 400 Marines to lift into an area where they suspect a division of North Vietnamese is. So they started prepping the zone the night before, put in 2,600 eight-inch rounds from a ship lying just off the coast. Then they pounded it with artillery. And then, just before the troops went in, they started running air strikes in there.

We had 11 helicopters which for the Marines is a lot. The Army will put several hundred helicopters into one spot, but we don't operate that way. It was really something to see when we lifted in those combat troops. We went down into the zone four birds at a time. Four, four, and then two. The 11th helicopter remained as SAR, Search and Rescue, in case somebody got shot down. It was beautiful. No one took fire; we just sat down in the zone and watched the troops pour out the back just like in the movies. It was a beautiful sunny day, not a cloud in the sky, and everything went like clockwork.

Afterwards we went out flying re-supply for other units in the field. The 164th is the king Squadron, no doubt about it. Our ground support personnel are just outstanding. In fact, we have one of the highest aircraft availabilities in the area. These people are really putting out, doing a good job and the pilots are working pretty hard too. Just an outstanding bunch of guys.

So all in all, I really enjoyed yesterday. I think I may have mentioned before that I'm getting a little more adjusted to what I'm doing over here. Also, I've gotten to the point that I'm really looking forward to making HAC. My roommate, Tom W., made it the other day. And I think I'm about ready. The flying has been just great lately, though some of our recon work has been getting worse. We had three emergency extractions yesterday that must have been pretty hairy, but I didn't fly any of them.

Well, young Bob W. just came home, so I'll close off for right now. Maybe later I'll get back to you.

September 18, 1968

Tomorrow morning I will have been in the country for three months. That leaves only 10 to go. A whole lot hasn't been happening around here lately. I've been flying some, but yesterday I was taken ill and was unable to fly for a day. I got sick during the night, and was still feeling under the weather the next day. Sick to my stomach and everything. Of course, the evening before I had been at the club, so I'll just let you draw your own conclusions from that.

I sent a package home today. I don't know if it will get home before this tape does. I guess that depends on how quickly I get the tape finished and in the mail. Sometimes its several days before I can get back to the tape recorder, but I just wanted you to know I haven't forgotten. I enclosed in my package three pairs of shoes, traditional Korean shoes that this Korean officer friend gave me, but I'm afraid they won't fit you. If they don't, just put them aside and I'll find a use for them later.

I went into the Ville today, the little Ville nestled around Marble Mountain, and got my first opportunity to see real Vietnamese at home. I got a little taste of their culture; very interesting. I went out with a Civil Affairs Officer, and a Gunnery Sergeant that works with him, me and another guy from our Squadron. We went along just because the guy was ordering some marble carvings for the Executive Officer of the Squadron. We had to go pick one out and tell them what we wanted engraved on it. While I was there I became completely fascinated by the marble work which is just absolutely beautiful, so I ordered one of those little name plates for a desk. If you want one, Pop, just say so and I will order one for you. I don't know if you want one with a Marine Corps emblem and Vietnam 68-69 on it, but if you do I'll be more than happy to get you one. Of course, I could get anything you want put on it, just a plain marble piece with your name if that is what you want.

That voice in the background is Tom making wise remarks, so we're just going to ignore him. Anyway, while I was there I saw the most beautiful piece of marble I've ever seen in my life and knew I just had to have it. I couldn't get the one I saw because it had been made up for someone else, but I ordered another one just like it, and for only $70. It's a marble tea set. It has a little tray carved out of solid marble, and some of the most intricate stone work you've ever seen. Just perfect. Then with the tray goes six or eight marble tea cups, I forget how many. And then, sitting in the middle, is this little marble tea pot. All of this had been carved out of solid marble, just amazing. I was amazed. It wasn't that heavy, but it was absolutely beautiful. So I'm going to get one of those and send it home. Also, I've decided to order a set of China. You might ask yourself, now what would the Ace ever want with a set of china, but I can get some outstanding chinaware over here. Well, now Weinberg wants to talk about China, says it's really made in Japan. Anyway, I think you can do without this, so I'll close off for now.

September 19, 1990

This morning I flew the first launch with young Tom as my HAC. It was really a good hop, everything went along real smooth. He does a good job and it's a lot of fun flying with somebody you've gone all the way through training with. This is the first time we've had the opportunity to fly together, but it seemed like the most natural thing in the world once we got up there.

Well, after discussing it with my roommates last night, I decided I might not buy the china I was talking about. They said it was uncouth. We talked about it half the night, which may be a measure of how dull our social life has become over here.

As Tom and I were walking home from the club last night, this jeep came by with two girls in it. Now, we've seen these two girls around eating in the club and all. Well, actually they're not girls. More like whores, really. But it really shot our morale down because here was this ugly old guy who works for Boeing-Vertol, a civilian even, these two honeys in his jeep. They weren't bad looking, had on swimsuits and everything, and the guy just whisked them off the base. We were really pissed. After a while without a woman, your morale really begins to suffer, I tell you.

Another thing we were talking about last night was getting adjusted back to the civilian world when we get home. It should be interesting to see how that works out.

Well, I've been thinking about writing some letters today. I do more thinking about that than actual writing. I just never seem to get around to it. Oh, by the way, that moustache wax you sent me works real well. I've got my moustache pretty much trained now. It's not doing too badly.

Right now Tom and I are engaged in one of our favorite sports – fly killing. These flies get wary. They'll start spiraling down into the zones after you kill a couple of them, but we're pretty sneaky. We'll wait until they land and then we'll zap 'em in the zone. So he kills one and then I get up with the fly swatter and I kill one. Lots of fun. Later.

Hi folks, this is John again. I just got back from the club and thought I would share my good fortune with you. I went over after dinner and decided to have a drink before I came back to the hootch. I had two dollars in my pocket so I decided to go into the slot machine room and throw it away. Well, I played for about an hour and came out with $25. It was something to do and proved quite lucrative.

Tomorrow I'm going back to the Ville at the base of Marble Mountain to pick up that marble tiger…or sculptured tiger, for our Executive Officer. He's leaving Vietnam on Saturday, so we've gotta get it to him tomorrow. I am having engraved on it: From the Men and Officers from HMM 164. Kind of a nice souvenir for him.

Well, I actually sat down and whipped off a letter tonight. I'll tell you, it's really getting tough to do because you get so stale over here. You live in the same hootch, get up in the mornings and fly, and the rest of the time we just lay around. We shower, go to the club, go to the shows, that's about the extent of our lives over here. You get stale and it's really hard to sit down and write a letter.

Or talk on a tape for that matter.

September 20, 1968

I got your tape today and sat right down to listen to it. It was a good tape and I was glad to hear all the voices, all except for the recording of that TV show with Phyllis Diller. Please, no more of that stuff.

Please tell everyone who spoke that I was glad to hear them, that all is going pretty well for me and all. I enjoyed listening to the sounds of the high school football game, and especially the band. Tom says you all are something else, you have great imaginations. He really enjoys these tapes too, almost as much as I do.

I wasn't on the flying schedule this morning. I had to go get that marble tiger I told you about. Then Tom and I went down to the beach for a swim, which worked out pretty well. I had been scheduled for a paratroop flight this afternoon but that was cancelled because the wind came up and they were afraid the guys might drift over into Indian country. They didn't want that. So I was all set to come back to my hootch when they collared me to fly a

ROK re-supply. Well, I've been flying a lot of those lately, I know most of their positions, so I readily agreed to it. Young Tom was flying wing and I was flying with the guy I had been with the day I got wounded. It was the first time I had flown with him since then and we were flying in aircraft 13! This is the third 13 we've had in the Squadron; the first two got banged up. The second one was the one on which my friend Ray D. got killed. Of course, I'm not superstitious, or anything.

Well, we went out and everything was going well. Then, all of a sudden, a controlling agency, one of those mysterious voices in the sky called Danang DASC, called us up and said they had an emergency MedEvac for us. We said okay, so we had to quit right in the middle of a ROK re-supply and pick up three recon people who had become ill and needed immediate medical attention. Well, by the time we finished that and went back and finished the ROK re-supply, it was about 5:30 and we were pretty tired. So we called Danang DASC. We always have to call them when we finish for the day and report that our mission is finished and we are going back to Marble Mountain. In fact, we have to tell them everything, when we leave and come back. They're the ones who keep track of who all is in the sky. So anyway, we call them and they said negative on our return to Marble, that they had three more emergency MedEvacs for us. Well, Roger on that, so off we went and picked up about 30 bodies altogether. Well, they had four KIAs on Tom's bird and we had four WIAs, and then we had 22, so there must have been 30 all told. We had on our bird, well, one of them died before we could get him back to the hospital. So we ended up with 25 WIAs and one KIA on board. They were all ARVNs, Vietnamese troops.

It's funny about the Vietnamese. I was down at the Industrialization Office the other morning and I happened to watch the Vietnamese workers come onto the base. Now when they come in, they have an ID card they have to exchange for another ID card that they keep while they are actually on the base. They trade it in at night and have to be searched on their way out. Anyway, they come to a window where there is room for one person to stand in front to change the ID card. But these people just mob the thing. It is just unbelievable, like a can of worms crawling allover each other. That's what these people remind me of, or perhaps a box of puppies. They just have no concept of getting in line and waiting their turn.

And the same thing applies when you take a bird into a zone, particularly one where there is some stuff going on. All of them want to be first and they just mob the first bird. Sometimes our crew chiefs have to be pretty brutal in their treatment of these people because they simply do not realize there are limits on what a helicopter can carry. Weight is pretty important to us, a thing we watch closely. And we just can't have these, well, I'm kind of at a loss for words. We tend to speak of them in colloquialisms, which makes it kind of hard to talk for your benefit back home.

Anyway, the gun bird that went with us warned us to be careful. The bird that had gone in this morning on emergency MedEvac had just been swamped with people and had trouble lifting out of the zone, had trouble getting people off the bird. A couple of weeks before we had a bird just swamped with Vietnamese to the point the crew chief was just kicking them off. Well, one guy managed to hang on until they were about 50 feet in the air and then he fell and was injured. So then they had to send another bird in as an emergency MedEvac to take him out. So you can see these people are real…well, they're real dummies, to be honest with you. I hate to say it, but it's one of the mildest things I can say about that. They are just…like animals.

We went into the zone to pick up our people. I wanted to just give you a little background before I told you about it. All the time we were flying ROK re-supply, it was only about three kilometers east of where all of this was taking place. All afternoon they had air strikes going on. They had F-4s and A-4s. They had some old A-1 Skyraiders and Huey gunships. There were 46s flying around. They had OV-10s flying around calling in artillery. And there was a Navy ship off the coast throwing in gunfire, so we had this stuff coming over our heads all day. We were trying to stay out of the impact areas; it was a mess.

But anyway, we could see all these explosions and watch all the air strikes and everything, so we knew there was something bad going on over there. Then we got the call to go in and pick those people up. It was exciting, but fortunately we didn't take any fire and got them back to the hospital all right.

That was about it. We came back, grabbed some dinner, and now I'm in the hootch. I'm sitting here not writing letters. I know I should be writing, but I just can't get up the energy. Tomorrow we have two strikes going, so that should be interesting. Right now I want to get some sleep.

September 21, 1968

It's a little bit after 6 p.m. and I thought to drop you a line before I head to the club. I went out this morning to help move 450 Marines from one place to another and finished up about one o'clock. We were a long time getting at it because some General came through and they ran us off for a while, like they were afraid for him to see us, or vice versa, or something. Finally they told us to launch. Then we discovered the unit we were going to move was in heavy contact in the pickup zone. So they had all kinds of Fixed Wing out there running strikes and zone preps and whatnot. So we headed out.

Now the only thing that bothered me was that the guy I was flying with seemed a little shaky in the cockpit. Have you ever watched a butterfly make an approach to a flower? Well, that's a good example of the approaches we made. I was more scared of him than of ground fire.

Oh, by the way, my roommate, young Tom, lost his cherry today. He took a round in his stub wing, his first one as a HAC. Congratulations, Tom.

"Thank you, thank you, well deserved."

Well, yes it was well deserved. Anyway, another funny thing about it was we had Fixed Wing on station running strikes and we wanted them to hold up for a minute so we could get this troop lift started. These air observers who call in the strikes are all grunts who've been converted to air observers, and they have no idea what is going on in the air. All they really know is ground operations. They don't seem to realize that helicopters need fuel and cannot orbit in a 360 for seven hours while the Fixed Wing run their preps and all that stuff.

So finally we told him: Look, Jack, we've got to run this strike in if we're going to get it in without having to go back to refuel. Well, he had one flight of Fixed Wing loaded with bombs yet to run and he was pissed that we wanted him to send them somewhere, else. So finally the guy decided to take his extra Fixed Wing and drop it right across the river from where we're putting our troops in. And that didn't help too much because in order to get into the zone we had to go almost through the flight path that the Fixed Wings were using on their drops there. So it just succeeded in screwing things up royally.

But to get back to the guy I was flying with…he was just scaring me to death. Finally he gave the stick to me and I did a better job than he had been doing. In fact, he told me as much when we got back. Of course, he is a Major so all I said was: Yessir. But I was glad to make it back in one piece.

Well, later on we're going to one of the other hootches to watch some flicks. That's something different. We don't get to see too many around here. Not more than every other day, anyhow.

So that's about it for tonight unless Tom has something to say. Tom, you got anything to say?

"I always have something to say. Hi, Mrs. Ace, Mr. Ace, and all the little Aces. I was flying with the Ace the other day and was fortunate enough to get my flight time I needed to make HAC before everybody else. Anyway, I flew with your son and I just wanted you to know he did an outstanding job. He's just a good shit in the air. He really knows how to fly and I really enjoy flying with him. The best part of the deal from my vantage point now is that I don't have to fly with assholes like the one Ace flew with today, though sometimes I fly on their wing. But it's just a great feeling to have the Ace at your side. I'm sure when he gets his 500 hours he'll be as fortunate as I was. Well, that's about it. It's Saturday night and here we sit."

Well, that was Tom's contribution for the evening. We'll go ahead and sign off now till tomorrow.

September 22, 1968

Well, today is Sunday and I just got back from flying a strike. We got hit with mortars last night, must have been 45 or 50 of them along the lines, the aircraft lines. We took two in our hangar…blew two gaping holes in the roof. One was a direct hit on my coffee mess. Destroyed two of my refrigerators and screwed up some other stuff. Also put holes in the West Virginia flag.

We were supposed to have a 10-plane strike today. We weren't sure we could even put 10 planes in the air, but the ground crews just do some outstanding work. Those guys really hump it, and they got 10 planes up. We were going to put in 300 Montagnards on the top of a mountain, the same place we put the recon teams in before that got shot up. It was as exciting as we thought it would be; got more birds shot up. I was lucky and came back unscathed. The zone was just terrible to try to land in. All they'd done was put an arc light in there and said: "...okay, fellows, there's your landing zone, go to it."

That is a hell of a way to build zones. It was on rocky terrain. There was scrub growth so you couldn't see what you were going to land on until you got down there. Then we were taking a lot of fire, and that sort of puts pressure on you. You know, you want to get down and get your troops out so you can get the hell out of there. Young Tom got shot up today, or at least his bird did. His co-pilot Steve H. was hit in the leg pretty badly. He's over in the hospital now. We're gonna go over this afternoon and see how he's getting along. He'll probably be MedEvacked out of the county because it's a bad wound.

Maybe later I can get Tom to tell you his story about the day's events. Right now I've got to get moving.

September 27, 1968

Well, you can see there has been some time lapse and I never did get Tom to tell his story. Right after I talked to you I went to see Steve. He was doped up pretty heavily, but had already gotten his wound taken care of. He had a cast on his leg, and the doctor said he'd be in the hospital from 8 to 10 months. But there doesn't appear to be any permanent damage and he shouldn't have a limp or anything, although it may take some bone grafts to put everything back the way it was. They took him to Japan on Tuesday and he is probably home by now.

The week has been fairly exciting for us. The day before yesterday, I went out on recon and used up most of my adrenalin supply for the week. I think that was the longest period of time I have stayed scared since I've been in country. It all started when we went out on recon and had five flip-flops and three extractions.

Well, the weather was pretty bad when we started that morning, so we were delayed. We finally got started on a couple of flip-flops. Everything was going well until we went into a place called Happy Valley. I was flying the lead bird. We had one team on board and went in to get the other team. Now, these guys had picked a pretty poor zone. We were in a hover at about 85 or 90 feet trying to drop into the zone, but we couldn't because there were several tall trees. One was a tall, thin hardwood sort of like a cedar. I was trying to watch on the left to keep us clear of the trees, and I could see the recon team on the ground. All of a sudden they threw up their weapons and began to fire. At that moment, both our 50 cal machine guns opened up. We use tracer rounds on these guns so you can follow your fire. I saw a burst of tracer go out and saw a tree just disintegrate. These 50 cals really do some work.

Anyway, we were taking fire from directly in front of us on the high ground and from our right-hand side. Also, we had an eight-man team on board and of course those guys just love to shoot. They love to kill z*ps, so they had their muzzles out the windows and were shooting like hell. Everybody was shooting so we pulled out. We called the guys on the ground and said "Look babes, you're gonna have to find a better zone if you want us to come get you." There was no way we could get into that zone, no way at all.

Well, we went ahead and did the rest of our recon while the Huey drivers pointed the ground people to another zone, and we told them we'd be back for them after lunch. Then when we got back to Marble Mountain and shut down we discovered we had totally destroyed four blades, three on the aft head and one on the fore head. We'd run them so badly that they had to shitcan them, then and there. There was no way to repair them. In truth, I was amazed those blades got us home, they were ripped up so badly. The hardwood tree was what did it. I think what happened was that when the shooting started, the HAC jumped a little bit and we dropped maybe two or three feet and hit the tree before he pulled back up.

Well, anyway we went back after lunch. The people on the ground had moved to another zone, but it was only about 300 meters from the one we had been shot out of. We tried to get in again, but once again they had picked a place that was just too bad to get into. So we went back up. In the meantime, the ceiling was pretty low and we had less than 1,000 feet of terrain clearance which put us well within the range of small arms fire from the ground. Fortunately, the zi**ers didn't try to shoot at us this time. The guys finally got into another zone we could work with and we finally got them out, but we were out there about 40 minutes and I was just scared shitless the whole time. I will tell you this, I put on about seven years. It's getting so bad my upper lip is going bald.

Last night we had another show at the club. It was nurse's night, and there were a lot of women there. They had a band, a lot of dancing and a General came to observe the frolic. Lots of fun.

This morning we went out to run a strike mission. We were going to lift about 400 Marines into this blocking position. They think there's a regiment or two of NVA in the area about five or six miles south of here. The intelligence people had gone out to check the area and didn't find anything, even though the enlisted people reported seeing some activity in the area. Well, they'd been doing the usual prep…bombarding it with eight-inch guns from a ship offshore, air strikes dropping napalm and 500-pound bombs. They just generally blew shit out of the place.

Well, originally the S-2 had planned for us to lift from the secure zone where we were picking up the troops and go low level into the drop zone. But there were some communications problems and the gunbirds called us to lift before the Fixed Wings had finished their last prep runs on the zone. We had two or three minutes worth. Then we started our descent right over the pickup zone, made a high speed, straight in descent, and then hooked into the zone. We were traveling south and had to hook around toward the west on our final approach heading. We got almost to the point where we started to turn and began taking fire. We were the number two bird in the zone with four other birds behind us. We were following the Colonel. He started taking fire just before he began his turn. We didn't notice until we got within about 100 feet of the ground that we were taking fire. Just before we received our first hit, our gunner opened up, the gunner on my side, which was the side the firing was coming from. I told the Major he had done a very skillful job of maneuvering me between him and the fire; it just happened to work out that way.

Anyway, our gunner saw an NVA jump out of a hootch and start running, so he killed the bastard with his 50 cal. I didn't see it, but the gunner said the guy just flew up in the air and dropped on the ground. So I promised to buy the gunner a beer. I'll buy anybody a beer who kills one of those guys, because I have developed an intense dislike for these worthless people over here. As far as I am concerned, the only good NVA or VC is a dead one.

We counted a probable kill for the guy. By the way, we took another hit in the cockpit today. It's funny, you know. You tune your ear for certain sounds. The Major didn't know we had taken a hit in the cockpit, but the instant it came in I knew there was a bullet at large in there. It's just one of those things you can feel. So I got on the ICS and said: Major, we took a hit in the cockpit. Then I commenced to look around and pretty soon I found it. It had come in above my head and from the rear, through the glass and into the overhead circuit breaker panel. It shot away our number one engine ignition circuit breaker switch there. It also shot away our rotor brake wiring and a couple of other things.

So we got back to the landing zone and got shut down. We couldn't put the rotor brake on, so we had to just let it windmill down. When we finally got everything shut down, we went to run the engines through a bit to cool them down to keep them from catching fire, as we always do. But we couldn't run one through because some of the wires had been shot away.

Anyway, we got to inspecting the bird and we'd taken a hit in one of the spars on one of the forward blades. That's a bad place to take a hit because it weakens the blade. Then we had two other hits in blade pockets, which doesn't make much difference. You can still fly.

Our skipper was a bit luckier than we were. He took a hit in one of his control cables. He was lucky to get back to the zone, because if that thing broke, he would have lost control. We had 16 people on each aircraft and weren't able to get them in the zone because we took so much fire. In the end they just moved the landing zone about 300 meters to the west and went ahead putting in the troops. But we had to shut down.

One of the Marines onboard the skipper's aircraft took a hit in the chest and was hurting pretty bad. They had to MedEvac him out. Also, the skipper's crew chief took a hit. But we wear these things called bullet bouncers. It's nothing more than a piece of armored steel enclosed in cloth. You strap it to your body and the crew wears one that has a back plate and a chest plate. The pilots wear one that just has a chest plate. But the crew chief took a hit on his bullet bouncer which is a testimonial that these things work because it saved his ass.

Well, they finally got the thing finished. We had to ride home in another bird and our birds were still out there in another zone. They had to send out maintenance teams to patch them up. They're going to external the skipper's bird back rather than try to fix it. They don't want to take a chance on someone crashing it trying to fly it back.

So this has been a pretty exciting day so far.

Now you know they've cut the R&Rs down to one per person per tour. But the powers that be decided that flight crews needed more because the day in and day out of going in there and getting shot at wears us down pretty fast. So they initiated a new policy. They don't call it R&R; they call it buying trips. Each Squadron sends one full crew a week. In other words, they send a pilot, a co-pilot, a gunner and a crew chief. A couple of times a month they send people.

Well, young Tom and I are going to Hong Kong on the 10th of October. That's before payday, so I'll need about $500. Make it $600. I want to get some clothes in Hong Kong, and some other stuff. The way I see it, I'm only going to be over here once and I may never again get to Hong Kong, so I want to make the most of this trip. I know I won't end up saving money, but what the hell, money is only money. I'll make it up when I get home. I'll try to find some things for you in Hong Kong, and next month I'll be back sending checks home on a regular basis as in the past.

Well, I can't think of anything else, so I'll just close out for now.

Chapter 10

Making HAC

Well hi folks. Today is the 28th of September, a Saturday afternoon. This morning we had a real tough mission flying causalities, guys who had been wounded, from the hospital ship back into Da Nang, so they could load them onto airplanes to be shipped back to the States. We finished that up early this morning and I got the rest of the day off.

Well, I just got your tape and listened to the first part of it. And I felt like I better stop and respond to some of what you said, Mom, before I listened to any more. The thing you have to realize is that we live in a completely different environment from anything you ever knew, and sometimes our sense of humor can become pretty morbid. I mean, after you've hauled some of these kids out of the field that are never going to be the same again, and after you've had some of your friends shot up, well, all I can say is that we joke about things that people in the real world would never joke about, death, for instance. You know, every time we go out on a hop, we sit around the ready room, particularly if it's going to be a bad one, and we make jokes. We say things like, well, what is it I've got that you guys want? Things like that. We just go ahead and start splitting up each other's stuff. But it's all just a product of our environment; it's not the way we really feel.

October 1, 1968

I don't know how many days it has been since I recorded last because I didn't play it back. I usually just pick up where I left off before. A lot has happened in the last few days. We went out on a strike Tuesday and got the airplane shot up again; took another round through the cockpit. It came in on my side of the glass, passed within about two inches of my head, and went into the overhead console.

One of my buddies wasn't so lucky. Well, he was truly lucky, in a way. He had a round fling off his hardhat. Now that will rattle your cage. As we were turning to go into this zone we realized in the midst of the turn that the zi**ers were just waiting for us. I was in the second bird as we went in together. Just when they opened up on us, the gunner on my side saw this NVA jump out of his hootch and start to run. So he opens up and nails the guy with a burst from his 50 cal. Now, I'm sure Pop is familiar with the 50 cal. I don't know how much you understand about this, Mom, but the 50 caliber is a pretty powerful weapon. It does some good work. When that burst hit that guy, it just knocked him head over heels. And then, when he hit the ground, the gunner opened up on him again, and was just rolling that guy across the ground like a tin can.

Now there was a time when I probably would have thought seeing something like that before my very eyes would, you know, make me kind of upset. But that particular day I thought it was the greatest sight in the world, and I would have given anything to be on the trigger. That little bastard and his friends still hidden around there were putting holes in our airplanes and, as I've said before, I've developed an intense dislike for these people.

Well, the action in our area has picked up quite a bit. I think partly it's because it's harvest time and the NVA and VC are moving into the area to get hold of some of the rice. So the action is stepping up.

Yesterday, we were supposed to put a strike in with 750 men, plus the Army was to haul in about 700 ARVNs. Meanwhile, another ARVN force was supposed to block the other end of the valley. Then the Marines and Army were to sweep down the valley and run the z*ps into a trap and kill them. Supposedly, there were about 1,500 z*ps, a regimental command post and a hospital and all that stuff. Of course, all of their stuff was built underground. There's no way they can build hospitals and headquarters above ground, not with our air superiority and all.

Well, these people had some pretty sturdy bunkers. The first two birds that started into the zone took heavy fire; the zi**ers opened up with their own 50 cals. We don't sweat small arms quite as much as we sweat 50 caliber fire which will just devastate an airplane. It's bad stuff. Well, the first birds into the zone were shot up so bad they couldn't lift out on their own, they had to go back and lift them out with those flying cranes.

But the grunts were dead-set on getting to this place, so this morning we went back in. In the meantime, all yesterday afternoon, they were running air strikes in the area. Then last night they sent in A-6s with 500 pound bombs and other various ordnance. And still later they put in an arc light so the B52s can come over at 40,000 feet and dump a whole stick of bombs, just saturate the area in an overlapping pattern. They just break a square mile into square grids and lay it waste. Any people who manage to live through it are reduced to vegetables. It's a good weapon and I'm sure glad they're on our side.

Anyway, they dropped the arc light in there last night about midnight. Usually, they come through with three B52s and just dump their load, and the g**ks never know they're coming because there's no noise like there is when Fixed Wing do bombing runs. The bombs just appear out of nowhere which has a psychological impact as well as the other.

Back to my story. They dropped them last night, as I said, about midnight. In between bombing runs, we ran more Fixed Wing attacks and artillery bombardments and then about 6:30 this morning. They dropped another arc light on them and went in with yet more air strikes. And then around mid-morning they did the whole show again to the point the smoke was rising about 4,000 feet or more. Way up there. And still they weren't done. They sent in about six more Fixed Wing attacks before we went in. That's 12 jets altogether. They were using napalm, 750 pound bombs, and daisy cutters. Now, a daisy cutter is, well, it's got a detonating device on the bomb which allows it to explode at ground level. You see, conventional bombs will bury 24 to 30 inches into the ground before they explode, which deflects most of the blast upwards. But a daisy cutter has a long probe on that begins detonation before the bomb actually hits the ground. A 750 pound daisy cutter will wreak some havoc, believe me.

Well, most of this was coming down on double-canopy jungle with trees about 80 feet high, just some of the densest stuff you ever saw. And by the time we got there, there was nothing left but dirt and crushed rock. I saw boulders that had been as big as a small truck now blown into gravel. It was just amazing.

And right after we started putting troops in; they were finding weapons and bodies allover. I haven't heard what all they found, but whatever was out there was in the hurt locker this morning.

Last night I went out on what was scheduled to be re-supply. Oh, that reminds me, I've got a story for you. It was in the morning and I was in the ready room when the operations Duty Officer asked me to watch the phone for him. Right after he left, the control tower called and said they had some round-eyed nurses who needed a ride out to An Hoa, and asked if we had a flight going that way. I said, Roger, that's affirmative. As it happened, I was scheduled to fly up to An Hoa in the afternoon. I told the guy I'd call back just as soon as we got aircraft assignments and let them know what airplane we would put the nurses on. He said okay and I didn't tell anyone except this one Major, a good guy who is going to Hong Kong with us. He thought it was a great idea. So when we got back from lunch and got our aircraft assignment, I called the tower back and said to put them on the aircraft, number 17 that we were flying. That we would be more than happy to run them out to An Hoa.

So we get out there and they're four Germans, two male doctors and two female nurses. Well, one of the nurses looked more like a boy than a girl, but the other one was the cutest little blonde you ever saw in your life. So immediately we invited her up into the cockpit where the Major was good enough to do the flying while I took off my helmet, talked to her, and got all her vital information and all. I told her the next time she comes into Da Hang to be sure and give me a call that I'd like to take her to the Officer's Club. Ever since which I've been telling my roommates I've fallen in love. Anyway, I hope to see her again.

The afternoon didn't go nearly so well. We ended up having to go to an emergency re-supply of a company in heavy contact in the field. This was another time I gained real appreciation for these 50 cals we carry around in our birds. We had to go in the zone and the first bird in took fire. He was getting cover from the Hueys which was pretty impressive to watch. But then we went in with our 50s blazing, which are guaranteed to keep their heads down, at least for a while and we managed to get in and out without taking any fire at all. I tell you, the sound of those 50s firing on your behalf is mighty comforting at times like that.

Oh, by the way, I got today from Virgil an invitation to his wedding, which kind of surprised me. I figured Virgil would be one of the holdouts, along with the Ace. But I guess he didn't, so I'll try to get off a congratulatory letter. Maybe I'll even pick the happy couple up a gift in Hong Kong.

I should be coming up for HAC check pretty soon. I don't need many more hours. I don't know, but I guess I'll handle it okay. I did well today. I flew most of the hop and the strike which was pretty hairy. I didn't crash or anything.

Another thing you mentioned was having to pay 20 cents postage due on the tape you got from me. Well, you can tell the Post Office somebody doesn't have their shit all in one bag, because we can send tapes and letters from here postage-free. Don't let them screw you like that. I'm over here striking blows for freedom and am not going to let them do that to me. Might even get riled up enough to write a letter to the Postmaster General, or whoever runs that place, and tell him how screwed up his organization is. Ah, well, I'm rambling. So I'll quit for now.

October 4, 1968

I just finished a 24-hour tour as Operations Duty Officer, which is not all that good a deal, to be honest with you. You just catch fire from everybody in the world. Everybody is shooting at you. I almost put on my bullet bouncer while I was up there behind the desk.

Well, I have news, I made HAC yesterday. I flew my HAC check into Thong Duc, the Special Forces camp Pop said he read about in the news, the place where they've been having all the trouble. I was flying with the Colonel, a really good man. Actually, Thong Duc is a dinky little place along a river in a big valley, and I don't know why anyone would want it. I've never been able to figure that out, but apparently the z*ps want it pretty badly.

I got a tape from my brother that worries me some. Dave seems all psyched up about coming over here. I'd like to talk to him and set him straight. I don't personally care for this place all that much. I'm sure you folks had something better in mind for me when you brought me up. And I'd just like to get the chance to sit down with Dave and tell him he's crazy. Of course, I know what it's like to want to come over here and prove something to yourself. I've been pretty proud of the way I've done, although I can't say I would have done it all if I'd had a choice. Of course, I was only flying co-pilot and didn't have much choice. But now I'm flying HAC and still have no choice because I've got a reputation in the Squadron to think of. It's amazing how things like that will spur you on, being scared shitless but at the same time knowing you have to do it. Maybe you're doing an extraction and you know if you were down there on the ground, you'd sure want those guys to come get me. You can't let these people down.

I'm really proud of the helicopter Squadron, to tell you the truth. I wouldn't want to fly jets or anything like that. I just...like helicopters. I like the mission we're performing. It's rewarding and satisfying. Of course, it is also dangerous as hell. In fact, other kinds of pilots are just amazed at us, mainly because we take getting shot up for granted. They really can't relate to that at all.

I've been keeping score, in case you're interested. I've had a total of 21 hits in the aircraft I've been in. Three of those were in the cockpit. I've had two aircraft shot up so badly that we had to put them down and leave them in the field until we could get maintenance teams out to repair them good enough to get them back home. I've had three of my passengers shot up. That's not a bad record for four months in-country, or almost four months. I just hope that I don't improve on it.

Speaking of which, I did something on my HAC check yesterday that was kind of an eye-opener. We were flying in a mission for the Marine Corps and ended up having it cancelled because the monsoons are coming in, and lower ceilings are beginning to obscure the tops of some of these mountains. These people wanted to go into a high zone and we just couldn't take them.

But we got diverted after the mission was cancelled to do a MedEvac. All the time we'd been working this mission, we could hear people talking about this MedEvac mission they had run. And they kept saying over and over it's only a priority. Well, that doesn't rate very high. It only means the guy needs to get to the hospital, but it's not that bad. Like when you're not feeling well and need to go on sick call.

But in the meantime it turned into an emergency, so we went out to pick these people up. It was some Vietnamese civilians and I don't believe I've seen anything quite so bloody in my entire life. Two men and a woman had been in a truck on Highway 1 which runs along the coast up toward the DMZ. It's a dirt road in many places; you've probably heard of it. Anyway, in one of the dirt stretches these people hit a mine. Now, I'd never seen the damage a mine can do, but I will tell you this was a decent size truck, and that mine completely tore away its front end. The wheels were gone and the motor block was blown out. The bumper was lying maybe 75 meters away. The cab was just demolished, and there were these three z*ps who had been in the cab.

Well, the first one they brought in, well, frankly I could not see how he was alive. He was, well, one of his legs were broken. The bone was sticking out of a hole. He was in a poncho and his leg was just flapping like a piece of rubber out the side. His arms were in the same state. His face looked like hamburger; he was just blood from head to foot. We got him in the airplane and I just kept looking around at him. You know, I'm sure he had to lose that leg that is assuming he lived.

The other guy was unconscious. I don't think he was dead, but he wasn't torn up quite as bad as the first guy. The woman was least hurt, or so it seemed to me, but she was in shock. Her skin was blackened, burned I guess, and she didn't have a lot to say. We hauled them off to the Vietnamese hospital and dropped them off.

Oh yeah, we lost a bird yesterday. Got the crew out okay, nobody was hurt, but it was one of these tactical blunders that makes you wonder about the idiots running this place over here. We had two birds out on recon and one was to do a flip-flop. Got into the zone and through pilot error he caught three aft blades in the trees and pretty much destroyed them, after which the aircraft was not flyable. But he did manage to pick up and set down in a creek bed that was right next to the zone. It was downhill about 100 meters from the zone in rugged terrain back in Indian country.

Well, the pilot, as soon as he got the engine shut off and everything, had his escape hatch pulled and was out. He left his helmet, his frequency card, his copy of the brevity code that the recon people use, even his pistol. He also left the co-pilot and the crew chief and the gunner to worry about the airplane. Well, the co-pilot was rather inexperienced so he decided to destroy the guns. That was his first thought and I have to say not all that bad a thought, but he didn't do it very well. We normally carry two thermite grenades in an aircraft for this specific purpose. A thermite grenade is pretty potent and will even burn through safes. They'd burn through that safe Pop has in his office, they really do a job. So he tried to pull a pin on one, but for some reason it didn't work. So he got the other one out, pulled the guns from their mounts and laid them across each other on the floor of the aircraft, and put the thermite grenade on top of them.

But the grenade rolled off the guns and burned a hole in the floor of the aircraft. So now we assume that the z*ps have two 50 cals with about 1,200 rounds of ammunition. They have all of our radios, the uniform of the UHF, the FM or what we call the fox mic. They have the IFF which is an electronic gadget that we use in conjunction with radar to identify friendly aircraft.

Well, I went back out there with the skipper just as soon as we heard about all this. I didn't really want to go because I knew it was a bad scene out there, but I'd been flying with him earlier, so I went. They had three recon teams in the area to provide security on the bird. They wanted to take in a maintenance team, fix it up, and fly it out. But when we got back out there, the recon teams were in heavy contact. That means they were in a hellacious firefight down there on the jungle floor. So the skipper decided he wasn't going to risk another bird and its crew to get the bird out, and he told them to go ahead and blow it up. Then he had to call Danang DASC which is a controlling agency to call the wing and get permission to blow this aircraft.

Well, a few minutes later DASC calls back and says we don't do that anymore. And the skipper says, well, you know, this bird's got two aviation 50 cals on it with 1,200 rounds of ammunition. It's got all the radios. It's got the IFF, the brevity code, frequency cards, and all that, and are we really going to just give all this stuff to Charley? Just let him have it? And the boys come back on the radio and just repeat that we don't blow aircraft anymore. So all the skipper could say was: "Well, its Charley's now, babes."

Well, today somebody finally realized the idiocy of that decision and sent two flights of Fixed Wing back out there with 750 pound bombs to destroy the aircraft. But as far as we're concerned they were too late because the zi**ers were in the area yesterday afternoon about 3 o'clock when this happened. So you know they went in there and stripped the guns, stripped the radios and all that stuff out of there. And you know they've got it now, there is no question in our minds that they have it. And I can only say it does not reflect very good head work on the part of the people that make the decisions around here.

That's about all the excitement for right now. I'll probably talk to you again tomorrow.

October 8, 1968

Well, today is not tomorrow, it's the eighth of October and quite a few exciting things have happened around here since I last spoke to you. You've probably read about the Special Forces camp at Thong Duc. In fact, I'm sure you have because Pop mentioned it in one of his letters to me. Anyway, the

NVA have taken over the place. Thong Duc was a provincial capital of some kind, a real dumpy little town. It makes Barboursville look like a real city, that's how dumpy it is. It's on a river and there's not much to it. Directly across the river is a high ridge line with a sheer cliff, maybe 800 feet high, facing the water. The NVA were firmly entrenched along there when we lifted about 200 Montagnards in there about two weeks ago. Well, they were mauled pretty badly; after a couple of days they decided to pull out and came down off the ridge.

But now the Marines have decided, in order to retake Thong Duc, we have to take that ridge. We believe there's a full NVA regiment up there with a hospital and everything. They say the ridge once had a coal mine and is honeycombed with tunnels that the NVA now use. So there they are on a cliff holed up in tunnels with machine guns and all they need. And our illustrious planners have decided we're gonna go in there and root them out.

There must be a special word somewhere for people who make these kinds of plans, but I don't know what it might be. Supposedly we were to lift in two battalions as blocking forces, one of which would be ARVN lifted in by the Army. Then we would lift in one of the blocking forces, after which we would put a Marine battalion right down on top of this mountain ridge where the Montagnards had been so badly used a few days before.

Well, we had once run an operation like this before and when we first went in we got shot at. So we put arc lights in and went back the next day and the g**ks were gone. The conclusion from this was that the arc lights made the g**ners go away. Thus, we arc lighted this ridge and headed in. Only this time the g**ks did not go away and shot the hell out of us. We ended up setting the Marines down along the base of the ridge where they were supposed to start working up.

Well, that night the planners got to thinking and decided we just had to get those people on that ridge, that we'd try again the next day. That night they went back in with more arc lights and maybe 750,000 pounds of bombs. Then we went back out to pick up the troops we were to carry in. As usual, the Hueys and Fixed Wing ran prep on the area, only this time the z*ps were shooting at the jets making the runs. They were shooting at the Hueys, too.

And then, just when we were about to go in, they shot down a Huey. So we cancelled again, came back home and decided to wait and see what happened next. That was in the morning.

They kept us around there a long time on stand-by. But that afternoon the group called up and said they wanted six crews to go out to Hill 55 as fast as possible because there was something going on. So we get out there and get briefed on this plan that must have come out of a funny book. We could not believe these guys were really serious about it. But here is what we did. We lifted from Hill 55 with six airplanes. They were all empty but we were hamming it up for the benefit of the g**ks who listen in on our radio. You know, This is one dash one, I got 16 passengers all set to go, and stuff like that.

In the meantime, the Hueys and the OV-10 were out conducting the regular zone prep just like always. We headed out to the area, got in a nice high orbit over Thong Duc at about 5,000 feet so they would know we were there. Then the skipper, who was flying in another 46 and was sort of the overall controller of this thing, discussed the route of approach with the guy who was leading the transports, and they got it all worked out.

I should note here that sometimes the g**ks will try to jam our fox mic using equipment they've captured. Sometimes they play music on it. Well, we could hear the music going in but as soon as the skipper started talking about the route of approach, the music stopped. Then when he quit talking, the music resumed. And then when they called for the transports to start in, the music changed to bugle calls and the sort of stuff they blow to sound the charge in cavalry movies. Interesting.

Anyway, we started down from 5,000 feet so they'd be sure to see us and then, at the last minute, we all broke away. We began flying parallel to the cliff with our 50s blazing away and all the z*ps down there shooting back at us. We were out of range of most of their fire, but it was pretty intensive and also pretty impressive.

Just as we broke away, the OV-10 that was controlling the air and everything called for fire and within moments about 250 eight-inch artillery rounds landed in the area. The first barrage must have been 50 rounds or so.

And then our leader called the skipper and said he thought he saw a route by which we could get back in there but I for one am not so keen on going back in there.

I'll tell you what's going to happen out there. The only way we're going to get those people out of there is to just go in with Marines and their M-16s and dig them out one at a time. I am convinced the z*ps will not give up that mountain easy. I mean after all those air strikes and artillery, they still won't let helicopters land. They have to have something big up there worth protecting. So the only way to do it is the old fashioned way, man to man. Our skipper says the only ordnance that will root them out is a hand-carried M-16 and I believe he is right.

One fortunate thing was that there were three Colonels in the skipper's bird, the very same ones who wanted to land on this mountain ridge. Now we keep telling these guys about some of the problems we encounter, but they never seem to quite get it. But when these Colonels saw the ground fire coming up at us, the muzzle flashes like Christmas tree lights along that ride, it was an eye-opener for them. I'm glad we had a chance to let them see.

Oh, by the way, I made HAC. I think I told you that before, but today I went out and made my first HAC hop. It was kind of interesting because the aircraft feels different on that side, you know. Your perspective seems to be a bit different. I think I'll catch on to it without too much problem and everything will work out. I think I'm going to do okay.

Well, I think I'm almost to the end of this tape, so I'll close off and get it in the mail. Check you later.

Chapter 11

Hong Kong

Well, hello family. This is Tuesday, the 15th day of October, and I have just returned from five days in Hong Kong. I can honestly tell you that Hong Kong is probably the nicest, friendliest town that I have ever seen, at least in the last four months. As long as you've got money in your pocket, every citizen of that town is your friend. So I've been away from the tape a while. Let me try to get caught up on the news. I think I told you I made HAC just before I left, and flew one HAC hop, maybe about seven-tenths of an hour. Then I went to Hong Kong and haven't flown since. We went to Hong Kong on a C-130. They were all flight crews so there were three officers and one enlisted man from our Squadron, some officers and enlisted from other Squadrons here at Marble, plus some guys from Phu Bai and Quảng Tri.

Immediately after we got through customs, we were inundated by all these Chinese guys pushing and pulling us to buses headed for the President Hotel. As soon as we got there, our baggage was whisked away and they took us to their shop to buy some clothes. You got to buy number one deal, Marine, I got a number one deal. I'm going to sell you handmade, custom-fitted suit, a number one deal at a special price for you, Marine, a special price.

So I got a few clothes; three suits, a sports coat, a pair of trousers to go with the coat, and a tuxedo. You may wonder why I bought a tuxedo. There aren't many occasions to wear a tuxedo in Barboursville, leastways not after you graduate from high school. Well, I don't know, it just seemed like the thing to do and I had the money in my pocket, so I said, yeah, I'll take one. And the Chinese guy said, well, why not?

It was kind of funny. The Chinese guys had a bit of trouble saying Ace. It came out ass or, well, mostly ass. At least I don't think they were doing that on purpose. Afterwards, this self-appointed guide took us to a restaurant and we had a Kobe steak. That proved to be our first good deal. It's Japanese beef. They take these cows, keep them in special pens, and feed them special foods with beer. Then women come in every day and massage the cows until they're big enough to butcher. And that has to be the tenderest meat, just outstanding meat. So that's how our tour got started.

Then our guide took us to our first Hong Kong bar. I tell you the whole town is just a real experience. You walk down the street and there must be a million bars everywhere you look. They all have guys hanging out in the doorways grabbing at you. Ahhhh, come in and see, we got number one show, number one, come in and see, such a deal. An odd thing happened while we were there. The guy who owned the tailor shop wanted to take a bunch of us out to dinner, a real Chinese dinner. We didn't actually see this guy, only the people who worked for him, and they were sort of seedy. So we decided this dinner deal was just another one of the number one come-ons we were getting from everybody. But we agreed to come and figured we had to go along, cause they said it would be free. So we showed up the first night and they said they had to postpone it, to come back tomorrow, and we said, well sure, Roger, and all that. The next night we went back on time and again they said they had to postpone it. To come back tomorrow, and we said, well sure, Roger, and all that. The next night we went back on time and again they said they had to postpone it. So by the next night Tom and I decided it just wasn't going to materialize, so we went out and had a great big wonderful steak dinner. And then we went to the meeting point again and what should we see but the owner of the shop, his wife, his son, his daughter, his son's fiancé, and various assorted other relatives. It was really a sincere thing they wanted to do, and here Tom and I had already eaten a big meal. Well, as you know, Chinese dinners are not the same as ours. This one had about 13 courses and you should have seen Tom and me trying to diplomatically eat small portions of each one.

Also, another funny thing, when we first got there, we actually got to attend a real Chinese fire drill. We got into the restaurant and there must have been about eight or ten of us plus all the Chinese family. I should say the people of Hong Kong are all Chinese, in case you didn't know. Anyway, the restaurant wasn't expecting us and did not have places reserved. But finally they found a room big enough for all of us. They had round tables with lazy susans in the middle. Everybody eats from a communal serving platter which you spoon stuff out of. Some onto their own plates and some directly into their mouths.

But no sooner did we get comfortable than they came in and said that room was reserved. So they put us in another room, and the same thing happened again. And then again. So we were all laughing like crazy about the Chinese fire drill we were attending. That was funny, but the dinner was real good. It was the first time I had ever eaten Chinese food, and you know I've always been squeamish about what I eat. I generally prefer cheeseburgers, you know, good old American food. Also, you would have been proud of the way I handled my chopsticks, like a real pro.

By the way, we bought our hootch mate an umbrella. It seems the big thing here in Vietnam is to have an umbrella, particularly the fold-up kind. You know, the handle on it works like a telescope and the umbrella folds. They collapse into about a foot long. We got her one of those and also she wanted some skivs and some bras. Well, we couldn't find her any bras, so we got her five pairs of skivs that will fit any size from zero to Jackie Gleason. She was real pleased to get that stuff.

Well, Tom and I ate in some excellent restaurants. It was kind of disappointing to come back to Marble Mountain and eat this hot dog goulash they throw at you. Kind of a real letdown for the old taste buds. We toured Hong Kong and I do have to say that the Chinese people have a lot more on the ball than the Vietnamese. Another thing we noticed is that the Chinese are very lean for the most part, at least in Hong Kong, and you don't see many fat women. Maybe the occasional portly gentleman, but for the most part they are very trim, and always cleanly dressed. They have beautiful hair, dark and shiny, a lot of body to it. We were quite impressed with what we saw of the Chinese people. I should add also that the ones who were not trying to get our money were very cordial to us and helped make our stay in Hong Kong most enjoyable. In fact I would like to go back there sometime if I ever get the chance.

Which is a lot more than I can say for this place. I must offer you this one story from Hong Kong. It seems we were going to entertain these Chinese ladies, elderly ladies. The one I was entertaining must have been about 45. At least Tom claimed she paid for the taxi with a Medicare check. Anyway, we were on the way up to our rooms there in the hotel when my date, for some reason, had words with the bellboy. And the next thing I know the two of them are squared off at each other in the lobby, and suddenly she hauls off and slugs the guy. It was beautiful and the bellboy, not to be outdone, rears back and dukes her right in the nose. So here's this woman and the bellboy, this elderly lady and this squatty little Chinese bellboy, duking it out in the hotel lobby. You just never get this kind of action in Barboursville. The next thing we know we have the management and security police swarming all over them and I decided to vacate the premises. We all just skated out and left her to her fate, though I have to say it was the best show we saw in Hong Kong. I truly would have loved to get that on film.

Well, one other story. One of the guys with us was set upon the minute we got to the hotel to by a tout, which is what the English hereabouts call a pimp. This pimp whispers in his ear that for only 1,000 Hong Kong dollars, he can have this beautiful Oriental jewel at his beck and call for a whole week. And our brave American pilot is fairly drooling as he lays across the money, and then, later on, in walks this ancient Chinese lady with gold teeth and the last time we saw our friend he was still trying to get his money back. I'm not sure what that's called, but I seem to recall something you told me once about buying a pig in a poke. By the way, I brought one of these Chinese suits back here with me and the others are to be sent to you. You should get them in about six weeks. I guess they send by sea or something.

Another good deal we could have done without was Hong Kong custom-made shoes. Tom and I got all dressed up and stepped out on the town wearing our new shoes, and by the time we got back our feet were just nubs. So if you see anyone headed toward Hong Kong, you best warn them not to buy any shoes there. It was actually kind of funny. I went in and asked to buy a pair of shoes and the guy asked if I wanted custom-made or ready to wear? He said he had a number one custom shoe to sell me. So I haggled with him about the price the way you're supposed to. Only later I found out I got the same price as for the ready to wear shoe, only it took a day longer. He whips out a piece of paper to get the outline of my foot, measures it and everything, writes some Chinese stuff on it. He says, "Ah so, come back tomorrow at 5 p.m. and we will have a number one custom Chinese shoe for you." Turned out to be a scam; he just took a box off the shelf for me. What an asshole.

So we were a little bit disappointed in Hong Kong shoes. Actually, there were several things about Hong Kong we found disappointing. The bar scene, for example, is something else. The Chinese have a funny custom when Americans walk into their bars. First off, a guy grabs you off the street and literally kicks you into his bar, and you just go screaming into the place. They want you in there so badly you can't believe it. And then the minute you come through the door about five girls jump on your back and ride you to the bar. Then the bartender starts throwing Hong Kong tea at you, wants you to buy it for the girls at only $10 a drink Hong Kong, which is like about $1.60 American. Anyway, a lot of guys in Hong Kong for the first time won't realize you don't have to buy all this stuff. Well, Tom and I would just shake them off. We almost had to carry flyswatters with us to keep them away. And then we'd tell the bartenders to keep their Hong Kong tea and pour us some drinks. Then we'd call for the mamasan. So we'd get her over and talk to her a while. And then after a while we'd say: "Mamasan, we don't like your girls, we want you." Well, this would really shake them up.

Now another strange custom they have relates to the men's room. You go on back to shake the dew off your lily and some lady back there hiding in the shadows hits you in the face with a wet towel and demands a dollar before you can go back for another drink. Well, what we'd do is walk by and flip the dollar into the room and watch them grovel for it. We spent one entire evening going from bar to bar shaking these people up. Really had a great time, and ended up not spending very much money. In a couple of places, the mamasan even bought us drinks to keep us there because she thought she really had a number one deal going.

Another thing we did was tell them we were officers sent to check the bars out before we sent our men to Hong Kong. This line generally got pretty good service from the mamasans. In fact, Tom and I are thinking about writing a book about how to get along in Hong Kong.

One rather amusing thing happened when we went up to look over into Red China. We had signed up for this tour with an English girl who ran a travel agency to go around the island. We were standing around trying to make time with her and all of a sudden we found ourselves on a bus headed for Red China, to the northwest, instead of into Hong Kong. Anyway, we'd been going through the countryside and all of a sudden Tom looks at me and says, "Do you smell what I smell?" And pretty soon there it was, a barrel from an outhouse. They were burning it just like they do in Vietnam. I've got to bring that concept back to West Virginia. We have to teach our fellow West Virginians not to dig holes for their outhouses, but to put all of it in a barrel and burn it once a week like they do over here. It really fosters a distinct neighborhood ambience.

But Tom and I were standing there looking across the water into Red China and this little girl comes up, pats my pocket to see if I have a wad of money in there and grabs my hand and starts yelling for me to give her some money. She wouldn't let go, just wouldn't. Followed me around for 10 minutes or more trying to get me to give her money. But I had already donated more than enough money to the people of Hong Kong, so that little girl had to go without.

We went out to dinner one night to a real fancy restaurant called the

Mongolian Room on top of a hotel. We got a table overlooking the harbor. It could have been a romantic setting had I been with a person of the opposite gender. Anyway, we had an outstanding meal of French pepper steak. And all during the meal, we had five Chinese guys hovering around the table. One handled the water bottle. When one of us took a sip of water, he immediately came to refill it, no matter how small a drink we took. Another guy did the same with the wine bottle. Had another guy on tap cutting bread for us. Real service like nothing I've ever encountered before.

So that's about all that happened in Hong Kong. Tonight we have a nurses' night show. I'm going to get dressed up in some of my Hong Kong custom made clothes and go over there to see what's going on. Tom's giving me the evil eye because he can't go, he's on duty tonight. But hey, what can I say? It could be worse, it could be me who's on duty tonight.

Oh, by the way, I don't have the coffee mess anymore. They had another shakeup in the Squadron infrastructure and now I'm Senior Flight Officer. In other words, I write the schedule. There's been four of us assigned to that job and since I'm the Senior Lieutenant of the bunch, they made me boss of it all. Such deal they give me, a number one deal.

Also, I'm an Operations Duty Officer. There are six of us on that job. But they took my coffee mess away and I am glad of it for as you know, I am not real hot on keeping records and stuff like that. I was actually having a little trouble with the coffee mess. My books always managed to balance but you would not have believed the stuff I went through to make them come out even.

Well, we're back trying to get in the swing of things again. Now I can only look forward to my trip to Sydney. I wish Tom could go with me but he has to go to Hawaii to meet his wife. Actually, this is a great joke. Tom wrote his mother-in-law and told her how nice it would be if she could come to Hawaii to visit with him and his wife and just as fast as she can reply she writes back and says she'd love to. So she's coming, or means to unless Tom can talk her out of it. Tom says it's a shame they can't get her on a C-5, because she's big as a tank. Says they had to paint a line on her head to tell if she was rolling or walking. So now he's trying to find a way to gracefully tell his mother-in-law he wasn't serious, that she doesn't really want to come to Hawaii to be with him and his wife.

Well, this tape is almost done and I want to finish it up and get it off to you so you can hear all about my trip to Hong Kong. It truly was an educational adventure. One thing you can say about being in Vietnam is you do have some wonderful opportunities to travel and see other parts of the world you wouldn't normally get to see. I am sure that in years to come, I will look back upon some of the trips I've made here with nostalgia. All in all, things aren't that bad here, not today anyway. The weather has been poor lately so we're not doing much flying. I'm hoping it will stay like that. In a way it's bad for me to be a brand new HAC just when the monsoons begin. I don't want to get rusty; I anticipate some hairy moments to come.

Chapter 12
Back to Work, Again

Well folks, here it is October 21, 1968, a Monday, and I just got back from a wonderful meal at the O club. I ran into one of my old buddies today and I've got him here in the hootch. I thought he'd like to say a few words to you.

"Hello, Mr. and Mrs. Astle. This is Roger H. You probably remember me from Pensacola. Ace and I used to live together down there. I just got back from Bangkok and it was tremendous, really tremendous. I spent five days and five hundred dollars there, and it was worth it. I had a good time. Well, I just wanted you to know that I'm taking care of your son here, making sure he stays out of trouble. I couldn't keep an eye on him in Hong Kong, but I guess he took care of himself. Nothing much else is new with me. I'm on board ship now, and Ace's Squadron will be replacing mine as soon as we leave in November. I've seen quite a bit of him lately, because our Squadron had been coming down here to Marble Mountain all the time. Well, we all sure wish we were home. I hope I get a chance to come up and see you all in good old Barboursville. That's about all. See you later."

Me again. Roger had his say and left. I went out and flew my third HAC hop in-country today. Everything worked out well. It was a re-supply mission. Didn't scare myself but only once, and I don't think I scared anyone else in the airplane because they didn't know what I knew. A lot of times you can feel things in the aircraft you wouldn't notice if you weren't driving it. But I tried to make a down wing takeoff with a fairly good-sized internal load and as I passed my ground cushion, I had not quite reached transitional lift. Since I was going downwind, it was even harder to get up enough forward air speed to pass into transitional lift. I started sinking out of the sky and had all my power on. I finally got out of it. Of course, I would never have hit the ground or anything. The worst that would have happened would have been for me to go back into my ground cushion again and hover along on that like an air taxi. But we didn't crash or anything.

Some of the zones we went into today had been taking fire, and a bird from one of the other Squadrons, had got the shit shot out of it yesterday. So we were expecting the worst today. But as it turned out, everything went well. We like to attribute that to pilot technique. I did enjoy the hop, and I had a real good co-pilot, one of the older guys. By that I mean he's been in the Squadron almost as long as I have. He's about ready to make HAC too. Flying with someone like that can be enjoyable, although I didn't share the landings or approaches with him. Like I said, this was only the third time I'd flown HAC in-country.

So that was about it today. It's a quarter after eight now, and I think I'll grab a shower and get some sleep.

October 23, 1968

I got a tape from you today and it's a good one. I really enjoyed it. I was especially pleased to hear Gigi's voice. It seems she has taken an active interest in my career over here, which is okay. I'm glad she's in the USO and suspect she will do well. I believe a lot of guys over here would be happy to see Gigi in the USO, or anywhere else for that matter. She could do some dancing.

Also, I enjoyed listening to Charlie C. You can tell him I will be looking forward to going out with him to do some shooting. Tell him I won't be kindly disposed to game wardens that try to confiscate guns after being over here where we solve our problems by shooting. You know, a VC pisses you off, you just kill him. Of course, I realize back in the real world, back in our world, you don't solve problems like that, or at least most of us don't. But it's always a temptation.

It's going to seem funny to go back to the real world and not be carrying my pistol around with me every place I go. I don't think I've had my pistol unloaded since I first got here. Every place you go, you carry your loaded pistol with you. You carry it to chow. You don't take it to the club but it goes everywhere else.

Well, things have been pretty quiet of late for our Squadron, for the last week or so at any rate. The weather has cleared up. Evidently that wasn't the monsoon season we were in. I don't know, but the last couple of days have been real nice. We got a lot of work done.

Yesterday I flew my fourth HAC hop in-country and came back unscathed. It's amazing. You begin to gain a little more confidence in yourself the more you get out there and fly. Oh yes, I should tell you what I hope to do tomorrow. You may recall a couple of weeks ago I said we'd carried some German nurses up to a place called An Hoa, and that I got a cute one up in the cockpit for some serious chit-chat. And also that the Major I was flying with was a good shit. He's one of the better Majors, and also an outstanding pilot. He and I talked to her a while, and I got her address and all. I told her I hoped she could be my guest sometime at the Marble Mountain O Club if she was ever in the Da Nang area. Well, we're starting to make some plans for the Marine Corps Birthday Ball which will take place, as you know, on November 10. And the big problem around here is getting dates. There is a scarcity of women over here, as you know. But for those who hustle come rewards, and of course those of us in aviation have certain advantages over our earth-bound comrades.

Anyway, I was talking to the XO today and I got the conversation onto the birthday ball, and I told him I had a problem. I said I know this girl and would like to take her. But the only way I know to get in touch with her is to just fly out there to where she is and land. And he says, oh hell, why don't you just go on out there. And then he said she could come in on the regular Mission 16 that flies between here and there and go back on it the next day. And if that doesn't work, he said, there will be so many helicopters flying over the I Corps area bringing in nurses and stuff that one more isn't going to make any difference. And so tomorrow I'm going to fly out there and invite her to be my date that is if I can get on a flight going in that direction. I will just land at her hospital and run in and ask her. I thought you'd like to know that not even a war can slow down the Marine Corps. I'll let you know how it goes.

Other than that, things have been pretty slow. We're getting ready to go out on the boat here pretty soon, and I'm looking forward to that. The only bad thing is they're going to take away our most experienced people when we go. The oldest guys in the Squadron will be the ones that came in the early summer, like me. So the Squadron will be looking to us to break the new guys in, if you can believe that.

Well, as soon as young Tom finishes polishing his boots over here, I think he'd like to step up to the mic and say a few words to you.

"Okay, we're all energized here now. Hi, Mrs. Ace. We just got through listening to that tape you sent, and it was really appreciated. A lot of good jokes and wisecracks. We enjoyed it. I know we've been harassing you folks back there in Barbersborough a bit. No, Barbersburg, right? You've really been putting up with a lot of nonsense from us. Well, I just want you to know we're not as bad as we seem. But I suspect you've probably guessed that already. Anyway, we're just trying to take care of each other over here and make the best of the situation. And a lot of times when we get off duty and take a few drinks, we get kind of rowdy. But we don't mean anything by it. Well, that's all I wanted to say."

I almost forgot to mention this last CARE package I got from you. I opened that today down in the squad room during one of the most crowded moments and everybody got a big charge when I started pulling out false faces and a rubber pumpkin. But when I pulled out the chastity belt with the ruler that really set them off. So that CARE package was enjoyed by everybody. Altogether, your CARE packages are pretty good. Tom just said he wants some more chocolate chip cookies, so you take it from there.

Now about Christmas presents. As you know, we move around quite a bit over here which means the more stuff we accumulate, the more we have to move it. So what I would prefer is for anyone with an urge to send me a gift just send it to you and let you keep it for me until I get home. Then I can have my Christmas in July. That's a real good plan because then I will only have to wait six months until my next Christmas, which will be a good deal. But if I start getting a bunch of stuff over here, it's just going to be a big job carting it all around with me. And then when I go home I'll have to cart it around again. So it would just be a lot easier if it was all there to begin with. Well, I'm going to close out now and go take a shower. Later.

Well, its' a little later in the evening and I didn't get my shower after all. A very good friend, Captain Shin, from the Korean Marine Corps, came in tonight and brought a plaque. It says Second Marine Brigade of the Republic of Korea Marine Corps. He had it engraved on the bottom. It says, To Captain John Astle, U.S.M.C. From Captain Shin. So he's here tonight and we've been shooting the breeze and I thought I'd get him to say a few words to you. So here's Captain Shin.

"Yes, thank you very much. Yes, my good friend is Lieutenant Astle. My name is Captain Shin, pilot. So I will stay here 11 months with very good, kind nice, 16 people, very good. I long time to remember. Before we had stayed here there were 12 people, 15 number one peoples. Oh, thank you. After five days I go homes. I long time remember my good friend Lieutenant Astle. Thank you."

October 26, 1968

Strangely enough, I was looking at some clocks today and discovered that the time here is the same time there in the day, even though we're one day ahead of you. Somehow that makes no sense. Maybe it's a 12-hour difference in the time. Like when its 5 p.m. here it's 5 a.m. there. Yeah, that must be it.

Well, I've been doing a lot of flying since I made HAC, mainly because we're short of HACs. And yet today, for some reason, I got off without flying at all. I don't know how I managed that.

Yesterday for the second day in a row I was flying re-supply missions that tend to drag out to all hours of the night, getting back after dark, two days in a row of this. I ended up having to eat in the enlisted mess, which isn't so good when it's hot and it's usually cold by the time we get there. Last night was the first time I had to make three approaches in unlighted zones after dark. I got in and out okay, but it was kind of tense there for a while.

Last night my leader lost an engine just before dark; had to set down at An Hoa. So I went in and picked him up and then we had some passengers to deliver, by which time it was after dark. Kind of interesting.

By the way, I don't know if I told you, but I lost the Stone out of my Marshall Class ring. So I got one of my buddies on his way to Hong Kong to take my ring along and have a black sapphire set in it. It's a pretty good sized stone and only cost $30, a good deal I guess. I kind of like it better than the green stone that was in there which I suspect was just a piece of glass.

Oh yeah, you know the German girl I told you about? Well, we dropped in the day before yesterday at her hospital out there. Well, naturally they thought we were bringing in MedEvacs and came running out with stretchers. But I waved them off and said that wasn't the case that I was looking for this girl named Sibene. So this German guy said she wasn't there, that she was over at another medical unit. I was having trouble getting directions from him, so finally he just said he'd go with us. So we took off and went over to land in this field there at An Hoa. The German guy wanted me to get out and go with him, but I said negative on that. No way am I going to leave my airplane and wander off through some ville.

So the guy left and a little bit later he came back with her. I told her what I had in mind, or at least some of what I had in mind. She said she'd like to go, but didn't know if she could get off that weekend. So I promised I'd drop back in as soon as I found out for sure what was going on. So yesterday I dropped out of the sky again into her hospital there, sat down, left it turning, left my co-pilot in the cockpit to watch the controls and everything, and went to find her. This time the German guy saw me coming and didn't have to ask what I wanted, he just went off to get her. She came out and I told her the thing would be on the 9th of November, some ceremonies and then a big party at the club. In fact, I think they're having a show that night.

So she checked her schedule and found she is scheduled to work that weekend. She gets every other weekend off. Then I discovered she's a lab technician, not a nurse, and she's the only one they have, so she can't trade with anyone else. But she said she would try to get into Da Nang next weekend and I might be able to see her then. I hope so. I would really be nice to talk to a round-eye girl, even if she does have a German accent.

I put on my Hong Kong custom made shoes today and I'm wearing them around to get them so they'll fit and won't hurt my feet. It's amazing. They're supposed to be made to the measurements of my feet, but I wore these things my first night in Hong Kong and they hurt something terrible. So I was thinking maybe if I just wore them around the hootch some and broke them in some more, they wouldn't hurt so much. I'd like to get them all broken in by the time I get to Sydney so I could wear all my spiffy Hong Kong clothes down there and see what I could get into. Right now that looks to be in January.

Oh by the way, I went down to do some more Christmas shopping at the PX in China Beach. I've got some stuff for you all, but I may not send it yet. I'm afraid the stuff might get messed up or lost in the Christmas rush. So I think I'll just sit on it for now, and you won't have anything from me under your tree this year.

As for me, there are two items I could use if you could send them. One is about six pairs of Burlington Gold Cup black socks. I know they cost $1.50 a pair, but so what? I've been wearing the same socks for the last week and a half because I'm short of socks. Actually, if you take them off and air them out every night, they're not too bad. But I could use some more.

And I could use some more of my BVD type skivvies. I've got a bunch of the baggy ones, but I sure hate to wear them. And most of my BVDs have seen their day. I could really use some more, and that's about all I need. And as I said before, you can just tell everyone who wants to send me something to leave it with you, please don't send anything over here.

As a matter of fact, I've probably already got more stuff here than I will be able to bring home, stuff like, sheets and pillow cases. I'll leave a lot of this stuff for the guys who follow me. I've got some stuff I've scrounged up that's probably been here three or four tours. In fact, a guy gave me six silver bullets that have been here for three tours that I know of, and are starting a fourth one now. They're supposed to be good luck, to keep you from getting shot down.

Speaking of that, my roommate Tom W., took some hits today. One of them hit his rudder pedal and knocked his foot off it. It didn't hurt him, but it scared him some. I imagine it would have scared me too.

So far I haven't been hit during any of my HAC hops. I took some fire yesterday or at least the guys on the ground said I took fire. My gunners didn't see any. Of course, I was so intent on my approach I probably wouldn't have noticed unless my gunners opened up. There's no way to keep from knowing when those 50s start banging away.

I shot a couple of interesting approaches yesterday, and on one of them I wasn't real sure I was going to make it. I couldn't get my bird into the zone. Had to do a couple of hairy whifferdills out there in the sky before I finally got it down. But I'm getting better. I'm still not what I would call good, but I'm getting better. I'm acquiring more confidence in my ability to handle the aircraft and make it do what I want it to.

One of the problems I had when I first got here and first started flying was setting up an approach into a zone so I could skirt all the areas where there was supposed to be enemy, or where they'd taken fire. You see, the Hueys get the brief for you from the guys on the ground who tell you where the friendlies are, where they last took fire, and then recommends the best approach. Then they use smoke to mark the zone, and you have to check the wind and see if you're light enough to make a downwind approach. If you're too heavy, you don't want to make a downwind approach because then you're going to have to plan your approach so you're into the wind. At the same time, you have to plan your approach so you won't pass over any areas where the bad guys are. Then you try to start your approach so that everything comes out finally at the right altitude and air speed. It's quite a challenge. I had a lot of trouble with it at first, but now I'm starting to get the hang of it. It's still not real good, as I said, but I'm getting better.

Oh, one more thing. My mail box has been empty the last five days. None of my women like me, I suppose. Oh well. I'll close for now and talk to you later.

Well, it's later on the same night here. The FRAGS came in for tomorrow and we've been working on the schedule. It's not my turn to write it, but we've only got 10 available HACs tomorrow, and we need all 10 to cover the FRAGs. I don't know what we'll do about the afternoon. We're really strapped because the other 46th Squadron here at Marble Mountain is going out to the boat tomorrow, and won't be doing any flying. We've got to cover for them, plus our own stand-by missions.

I am scheduled for my first shit sandwich as HAC tomorrow. I've got this stand-by mission where you go out to one of the zones, sit down and wait. The last three times they've gone out to that area they've taken fire. I think I told you earlier about Tom taking a round on his rudder pedal; that was on this mission. It should be exciting, to say the least.

It's 11:30 now, so I'll sign off. I've got to get up at 5 a.m. Tomorrow night I'll probably have some good stories after an exciting day of combat flying. I'll talk to you tomorrow.

November 2, 1968

Well, it's been several days since tomorrow. Just thought I would quiet your fears. That shit sandwich I was worried about turned out okay. We didn't take any fire or anything.

We went back out the next day. My airplane went down into the zone there and I was starting it up again as we getting ready to go out on a troop lift when I blew a hydraulic line. So I had to stay there most of the day until they could bring a replacement part out and fix it.

That afternoon we were sent out on an emergency extract for a recon team, only it turned out we were to put in a reaction force to link up with them, not pull them out. Reinforcements, you know. Then we winded up doing some MedEvacs out of the area, kind of a bad day. I got about an hour and a half of nighttime flying. It was pretty shitty. I don't care for night work. It really drains you.

In fact, I had to go out tonight at about 5:30 or 5:45. I was sitting in the club having a drink and all of a sudden the SDO came up and said, "Hey Ace, you've got to go flying." I said what? And he said, "Yeah, you've got an emergency re-supply." So we went out and did that. I don't like landing in jungles at night. I don't like it at all. Night flying is bad because you lose your depth perception, and it's tiring because you're a lot more tense. You have to be on your toes at night even more than in the daytime. Well, I had about an hour and a half of that tonight and I am beat, to be honest with you. I didn't have to fly today, though, and I spent most of it making myself an ashtray out of an artillery shell. People look at me funny when they see me dragging in these souvenirs, but I thought it would be kind of nice, so I got one. It's brass and will look great when it's all polished up.

But nothing else much is happening. The peace talks haven't affected us much. The bombing halt hasn't affected us at all. I don't think anyone has told the people down here in the south what is going on. They're expecting us to get hit tomorrow night, probably with a rocket attack. The VC need to do it since there's an election coming up. They need to make their presence known, but I don't know if they'll accomplish anything.

There was a funny thing the other day when we were out flying. We heard this guy come up on the radio and say, "Hey, I just found some mortar tubes. They're all set up in this village square." Then we heard this other voice say, "Well, I can't because they're in a friendly ville."

How about that, huh? Right in the middle of a friendly ville they got mortar rocket tubes aimed right at Marble Mountain. You can't go in there shooting because you might piss off some friendly civilians. And you can't do that because that is bad for the war. Well, I've got my own ideas about that. I say if the friendly civilians are going to let people put rocket tubes in their village, then as far as I am concerned, they are not friendly. And the best thing for us to do is just go in and blow the shit out of the whole ville. But then I guess I am a reactionary or a warmonger or whatever term they use for people like me back in the States. I'd feel a whole lot better about this war over here, if they'd just let us go out and do the job regardless of who got hurt. Maybe that sounds harsh, but a lot of Americans are getting hurt because we don't want to piss off the Vietnamese. And the Americans are a lot more important to me than the Vietnamese.

The other day I was out in the zone when they brought in a detainee, a Viet Cong suspect. It was an old woman, so wizened up I couldn't believe it. She was on a stretcher all curled up in a ball. She was wearing a blindfold when they dragged her in. It's hard to know how to think about enemies like that.

Well, I think I'm going to close off and go take a shower and go to bed and get some sleep because I think I earned it today. I didn't fly all day but flying at night more than made up for it.

November 7, 1968

Well, today is the seventh. I can't remember when I talked to you last. Probably been a while. Last couple days here haven't been all that good. We got one unit, the 26th Marines that have run into worlds of shit. There's been a lot of trouble with the MedEvacs getting the wounded out. They're up in the mountains north of Da Nang, maybe five or six miles, and pretty close. The people up there have real problems. We went in there about three days ago, four days, I don't know. I was flying wing on this guy who went in to get a MedEvac. On the first pass he took intense fire, heavy automatic weapons fire, took some hits, and just pulled out. The Hueys put some fire in, ran Fixed Wing gun runs, the usual. Then he went back in again, took more fire, more hits, and pulled back out. So he decided to try one more time. He went in a third time and just got the shit shot out of his airplane.

If I sound strange just now it's because I'm trying to take off my boots and talk at the same time. Anyway, I heard him going in and said in his last transmission he was taking fire. The next thing I saw him dropping down and I was pretty sure he was in trouble. I was pretty sure his radios were out because I couldn't raise him. So I fell in behind him and we headed back to Marble. I was flying on his wing. When we got close to Marble, I radioed ahead to the tower to have them clear a pattern and get emergency crews on deck. I told them I had no idea what his intentions were since we couldn't talk. Well, right beside Marble Mountain is a hospital with a landing pad so he decided to go in there first and drop off his crew chief who had taken a bullet in the chest. So I had to call the tower and tell them all about that.

We went back that same afternoon. I was flying wing on another guy. We went back to the same place and managed to get the MedEvacs out with no problem. We didn't take any fire. Then yesterday, we had a bird go back in there for a MedEvac, the same place, and get shot up. This morning we had yet another bird go in for the MedEvac and get shot up. And we had an airplane out there again this afternoon trying to get that same MedEvac out, and as a result we got another crew chief wounded, and another plane shot up.

One of the gunners was cited for outstanding work in this thing. They took a hit in the transmission, the mix box, and this guy went back there, climbed up on the edge of the ramp so that he's looking down at the ground, and stuck his fingers in the bullet holes. Now the oil in the transmission is pretty hot and he just literally cooked the ends off two of his fingers. But they made it back to Marble Mountain, got the crew chief to the hospital and there is no way they could have flown back without oil in the transmission. Just no way. They would have had to put down in the field. So they're gonna write this kid up for the Distinguished Flying Cross and he deserves it. No question, he never hesitated. As soon as he knew they were losing oil from the transmission, he just ran back there and did what he had to do to stop the leak. I tell you, when you've got kids like that working for you, you can't go wrong. They are some of the finest people I've ever seen, love to get into the action. We pilots may be a little hesitant, but the crew chiefs and gunners just love to get in there. They just work their balls off. These guys, a lot of them anyway, are still out there tonight working on their airplanes. They'll get to bed maybe at midnight and be back out there at 4 a.m. to get their birds pre-flighted for the morning launch. Then they'll be out flying all day eating C rations. Sometimes they even sleep in their planes. Some of them have even extended, serving second and third tours over here. They are simply outstanding people to work with; I can't say enough about them.

Well, that's about all the excitement we've been having. I don't know if you've read anything about it in the papers. I've been real busy flying a lot. Flew all day yesterday and much of the night too. I went out to the carrier to qualify and didn't get back until 9 p.m. But I'm both day and night qualified on the carrier. So I guess the Squadron will be deploying here the end of next month aboard ship, and word has it we'll be aboard for six months. If so, that will be a good deal. You get a little slack on the ship and don't have to support the whole world like we're doing now, all these weird messes we get into. Your work off a ship means you support the landing forces off the ship and that's pretty much it. You send out four birds a day to work with them, and that really isn't so bad, not bad at all. Well, I'm scheduled for recon in the morning, and I need to get some sleep. I don't really like recon a whole lot. It's probably the most frightening mission that we fly. I don't mind strikes or any of that even though they're probably more hazardous than recon. On strikes you have all your friends around; on recon you normally go with two birds and only one goes down into the zone. Of course, you have two Huey gunbirds covering you all the time, but it still gets kind of hairy sometimes. I'm not really looking forward to it. On the other hand, it is raining. Maybe it will rain us out. I hope so. All for now.

Chapter 13
The Nuts and Bolts

Well, hi folks. Today is 12 November, late in the evening. We went out this morning to fly as part of a four-bird troop lift, went out into the zone, shut down to get a brief but when I went to start up again I couldn't get my number two engine started. Our ignition was functioning but our actuator wasn't programming to start so it wasn't getting any fuel. I had to wait for a maintenance team to come out. In the meantime, our crew chief got to screwing around with it that time the mission was completed, so I didn't really get a whole lot accomplished this morning.

I went back out this afternoon. We hauled some troops around and then the weather got shitty and we had to come home. That was about it; not a very exciting day at all. Of course, you never know when you launch what's going to happen.

I was going to say something about the tape that I got. I enjoyed it quite a bit; it had a good variety of stuff on it. I didn't hear too much from Pop, except for his jokes. Some of the guys were in here when he came on and they got a big kick out of his jokes. So anytime you all can come up with some jokes, rest assured we will be more than happy to laugh at them.

Also, you can tell Tom T. I know 45 minutes is point-seven of an hour. We have a little chart in the line shack to help us figure things like that out, how many minutes is equal to a point.

"Hi home." That was Jim A. You probably don't know him, but he and I went to OCS together. Oh yeah, tell Tom Thornburg I'm glad to hear he's now a flight check engineer. Also tell him I'm flying as an Aircraft Commander now. I've got almost 50 hours of HAC time and am enjoying it immensely. Its added responsibility, but I think I'm handling it pretty well.

I heard some scuttlebutt on myself today. I was talking to some guys from another Squadron who said I have the reputation for being one of the most cautious pilots in the Squadron. But that's the way I operate. I figure it's better to be cautious and stay alive than be one of these hot rods and push your luck.

You can also tell Tom I made Captain. I was in the group communications shack when the orders came in. I've been waiting for it, and I was down there waiting to see if my name was on the list. So I knew the day it came out, but I don't know yet when I'm going to get picked up. Tom said he didn't understand why a person whose name was on the list didn't automatically get promoted. But the way it operates is, you get selected and then you have to wait for a slot to open up before you get the actual promotion.

You asked how the election was affecting us. So far not much. I really couldn't have cared less because I don't think it's going to have that much effect on what's going on over here. And as far as the bombing pause is concerned, we don't really think it is accomplishing anything other than giving the NVA a chance to replenish their supplies. We've been hurting them pretty bad here in the Da Nang area, keeping up steady pressure, finding their rice and ammo caches. Now with the pause in the bombing they'll have opportunity to replenish those caches.

Tet could be big again this year like last. I don't know, it's hard to say. One thing I do know. Even the day after the bombing pause began it didn't affect us here a whole lot. I don't think anyone has gotten the word to these people around here; they're still shooting at us with real bullets like they were before. And, well, I don't know.

Well, I guess I should get some rack time. I've got a full day ahead tomorrow, a re-supply mission in the morning and then again tomorrow afternoon. So I should have a full day. Oh yeah, I'll want you to take my motorcycle into the shop and have it tuned up for me before I come home. And have them put some preservative in the cylinders, clean the oil out and all. I'd sure appreciate it.

I'm sorry for the delay in the tape, but they've been working us pretty hard lately. You come in so tired that all you want to do is go to bed. You don't want to sit around and talk into a tape recorder. But I'll try to do better, maybe a tape a week like I promised initially.

Also, Mom, as far as your students at school are concerned, I know they would like to know what it's really like over here, but it's really hard to convey. You know, we tell our stories and all, but the emotional context doesn't really go into the tape at all. A lot is lost in translation. This would be difficult for anyone to understand who hasn't experienced it. But I guess we can try a little harder.

Well, I'm going to close off for now. I'll try to talk to you again tomorrow.

November 20, 1968

Boy, what a hell of a day this has been. This morning we took part in what was probably one of the largest coordinated operations they've had in Vietnam. We had almost 70 aircraft involved lifting 2,700 Marines into a ring around some g**ners out there in the same area we've been flying in ever since I got here. We went out at 4 a.m. to get briefed, get our birds and get out to An Hoa. Last night we were hit by mortars that seriously wounded two of our guys on the line. One of them was a Gunnery Sergeant, a pretty good shit. The other guy was one of the mechanics. Then we had about seven other guys take shrapnel wounds, none serious. Probably no worse than the one I got. So that's how the day started and it must have been an omen of things to come.

So we went on out to An Hoa and got all started to do this troop lift. I guess the most frightening thing from my point during the troop lift was that we had

14 aircraft nose to tail spread all over the sky. I've never seen that many working together. I've seen as many as 11, but never 14. But anyway, we started our approach into the zone and there was a cloud layer about 300 feet off the ground we had to penetrate. So as we started down we broke off. There was one section of six and two of four. Anyway I started down and when I punched through the clouds I looked over to my 2 o'clock position and saw three other birds coming around right at me, another one entering a pattern in the opposite direction, and they all had their 50s blazing. I told my gunners to cease firing else they might shoot up our own people. They hadn't seen them yet because they were so intent on shooting down at the ground, you know, at any possible positions we might take fire from to try to neutralize them.

So we got the first wave in okay, and went back and got the second. It was then the shit sort of hit the fan. "Excuse me." We heard someone come up on his radio and you could tell by the tone of his voice he was in trouble. He was just screaming that he was taking fire. Then we heard him trying to talk to his gunner or his crew chief. At this point we knew there was some bad shit going on because the switch on the stick that you use for the ICS and the radios is a two-position switch. Well, the first position is for the ICS and then you squeeze a bit harder and that's the radio.

Well, this guy was talking to his crew chief on the radios so we knew immediately he had a death grip on the stick. And then he was screaming that he'd been hit in both legs, wanted his crew chief to get up there with a bandage. The co-pilot took the aircraft, waved it off, and flew him to a hospital. He only took one hit.

Another bird from the same Squadron went into a zone, caught a blade on a tree, and ended up staying there. But most tragic of all was a bird from HMM-265 that went into a zone and just exploded as it touched down. The HAC was a friend of mine, a guy I knew quite well from Pensacola. Only one crew member got out, the crew chief, and he was in pretty bad shape. He was burnt pretty badly. There were some survivors among the 12 grunts who were on board. But when he hit the ground and exploded, the aircraft just rolled over and started to disintegrate. Anyway, I've had, well, I'm sort of losing my train of thought here.

Anyway, when the aircraft began to disintegrate it wiped out the staff of one of the grunt companies, killed the company commander, the XO and the company gunny. Killed the company commander from another company that was in the bird right in front of them, and wiped out a whole bunch of grunts. Then they had another problem with all these wounded on the ground, many of them just burned horribly, and they had to get a MedEvac in. But the fire was just unreal. Five birds tried to go in and all five were shot out. There was just no way they could get in. Finally, one of our Majors, a guy who will just throw his nuts out the window, who will try anything, got down there. Got to hand it to the guy, he does a good job. Well, he got in there and got all the MedEvacs out and took them to the hospital. He had to leave the dead, didn't have room for them. We haven't found out how many of them survived. I hope they all did.

Later on in the afternoon I was out at An Hoa. We were shut down at the runway there and I saw the guys casually bagging the bodies. It just leaves me kind of cold, I'll tell you, seeing stuff like that. Well…

Anyway, our Squadron was lucky. We didn't take any fire. No hits. We came out of it unscathed except for the casualties we got with the mortar attack last night. I didn't get to finish the strike because my bird went down and I had to return to Marble. It sprung a hydraulic leak in one of my control boost actuators. So I had to get the bird back as fast as I could.

Anyway, I had to go out and fly recon, which is kind of a bad deal, and I was already tired. Well, you might wonder at that, how sitting up there in the cockpit could be so tiresome. Well, I'll tell you what. I've worked some strenuous jobs in my day, but I have never done anything that drains you like flying when you're scared. It just drains you physically and mentally. It just completely exhausts you.

And as if this morning weren't enough, they had me back out this afternoon on recon, an emergency extract of a recon team. True to form, they were in contact when we got there, exchanging hand grenades with the g**ners. But we managed to get them out.

Then when we got back, and had about six hours flying time in, they tried to divert us to go out on another extraction of a recon team that had got into trouble. Fortunately, my leader bird was down and there wasn't time enough for us to get another one and go back out again. So they found someone else to do it.

So here I am in my room. I think I'm just going to call it quits and go to bed and just hope I never wake up to another day like today. I don't know how long it's been since I last recorded a message to you, quite a while I imagine. I played back the end of this and heard myself say I would talk to you tomorrow. But I have the feeling that was close to a week ago. Anyway, you can tell your friends young Ace took part in Operation Meade River. I'm sure by the time you get this tape you will have heard about it on Huntley-Brinkley. Chances are you have seen some Yankee tango birds because we had an NBC cameraman onboard one of the aircrafts this morning. Well, I'm going to knock off and talk to you again, well, whenever.

November 28, 1968

Hi folks. Here I am again. Not a lot has happened since the 20th when we flew the big strike which was apparently the last day I spoke to you. Well, you wanted some stuff you could play for your students, and something about how we feel about the Vietnamese. That's a hard one. We're not really involved with the Vietnamese, with the exception of our hootchmaids, we're pretty much isolated from them. And the hootchmaids are so intent on gleaning all the money they can from us that they're really not interested in anything else. So we've never really had a chance to discuss with them how they feel about the war. They really don't speak much English anyway, and we don't speak any Vietnamese. So I really don't have anything to offer you on that score.

We're going to be moving pretty soon, probably about the 29th of December. We're going on the boat. Then with typical Marine Corps planning we'll be there for a total of 96 hours and then off-load the Squadron again at Phu Bai. What we can't figure out is why they want to put us on the boat for 96 hours. It's a hell of a lot of work involved in moving an entire Squadron with everything, you know, all the gear, all the papers and files, all the machinery, all the tools and spare parts, the whole works. It's a lot of work moving that along with everybody's gear aboard the boat and then turning right around 96 hours later and doing it again. We haven't really been able to figure out how our brilliant planners managed to come up with such a brilliant scheme.

After we come off the ship, it will go for Christmas leave in Okinawa. Then it will go from there to the Philippines, pick up a new battalion there, train them there, and then come back to Vietnam. In the meantime we'll be working out of Phu Bai, which is not a very nice place to live. The conditions are on a lower scale than what we're used to here at Marble Mountain. Actually, Marble has probably the finest living conditions of any place in the Corps area. But alas, there is a war going on and we cannot enjoy the luxuries forever. I suppose we must steel ourselves to the dim prospects of Phu Bai.

Now once the 20 days are up and we go back aboard the boat, we're not sure how long we will stay. There's talk we may stay for two months, and then there's also talk about keeping us on for six months. I personally would prefer the six-month stint because that would mean I would be coming off the boat to go home. Life on the boat isn't all that bad. You get used to rocking all the time. The quarters are nice. The food is pretty good. The boat is clean and I have lots of slack time. You don't fly recon and you don't have to stand the Bald Eagle. All in all, it's not so bad a deal.

I got your package the other day, actually three boxes loaded with goodies. I'd like to thank everyone involved. The puzzles were especially appreciated. We're getting plans made for a Christmas party here in our hootch. I've gotten several gifts, a couple of little things from Uncle Don and Aunt Esther. I believe that's who it was. Then today I got a package from Ann B. She included a Christmas stocking with, listen to this, a can of deodorant and a tube of toothpaste. I suppose that's a way of saying my pits are smelly and I have bad breath. Which is true enough, but how did she know? Also, I got from her tins of cookies and cashew nuts. We've been stockpiling all the food you people have been sending to us, and we're going to bring it all out for a big party. Bob got a small Christmas tree from home, the kind you assemble. We also have some booze, Christmas spirits, as it were. I think we'll have a pretty good time.

We've grown quite close, Tom, Bob and I. I don't think any of us could have asked for better roommates. It seems like each of us is willing to put out about 75 percent towards getting along with everybody else, so it's worked out real well. We've never had any arguments among ourselves or anything. The best times I've had here at Marble Mountain have been with these guys.

I recorded a special tape and sent it off to Gene C. He had sent me one from his radio thing. He's on a radio station in Annapolis, Maryland. Anyhow, he sent me one of his tapes and I thought he might enjoy listening to one of mine. A lot of the guys added messages to him.

Tomorrow we've got another strike going, a 12-plane strike. I'll be working as the Operations Duty Officer, so I'll be sitting on the sidelines monitoring everything on the fox mic.

By the way, the new darts you sent along are great, much better than the earlier ones, and we've been using them a lot. And the puzzles. I think I said that already. It's strange how we zip back and forth from times of great excitement to hours of impenetrable boredom. One of our friends in a nearby hootch is now playing a tape, about 50 feet long, of farts he has recorded over the last month or so. Nothing but farts. Every time he feels one coming, he runs back to his hootch to get it on tape. It sounds like something the old Ace would do, but I just didn't think of it. But he's going to bring it over for us to listen to. We can hardly wait.

Now if you can hear that right there, that is a sound that strikes terror into the hearts of helicopter pilots, launch the Bald Eagle. The Bald Eagle, as you know, is a stand-by crew for any emergency launch that might happen to take place during the night. Tonight they're going down into an area just south of a place called Hill 55 into where they're having an operation, Meade River, at this time. It seems that a unit down there has run out of small arms ammo and needs more. Also, it's a rather bad area because there are known 50 cal positions, and the unit that needs the re-supply of ammo has been taking 50 cal fire which is not all that good.

Okay, now, wait a minute. We just got a reprieve. We just got a call from group and the mission has been cancelled until further notice. And the reason, believe it or not, a reason we don't often hear, is that the area is considered just too hot for helicopters. If you can believe that.

Okay, if this sounds kind of convoluted it's because I've been turning the recorder on and off. We just came through a very real exercise that almost sent me flying around in the jungle at night getting shot at. We are now sitting here talking about how happy we are that we didn't have to go.

Tom was just saying that we'd go if we had to go, but the prospects aren't all that good and anyway, he doubts if the grunts out there are really that low on ammo. They generally call for re-supply well before they run out, and they can see the situation as well as anyone. We could always get called again later if they're really hurting out there. You can just never tell at this stage of the game whether we will have to go. If we do go, we'll take a full crew, two gunners and a chief. You'd want all the firepower you can muster out there. The thing is that if the crew chief is going to be in the back working the external load, there is no way he can also work one of the guns. So you need two gunners.

Well, it looks like I may have dodged a bullet, but I won't know until morning. I'll check back in with you later.

November 29, 1968

Well, I just wanted to check in and reassure you we didn't have to launch that Bald Eagle we were talking about. They just called it off, didn't really need it, which was a good deal. The weather was not all that good, and neither was the enemy situation.

Well, we're expecting some action shortly, and I have to go to work. So I'll get back to you when I can.

Later....

Okay, I'd like to answer the questions Dan Mc. had on your last tape about Pensacola and the flight program and Marine aviation in general.

The first question related to living conditions in Pensacola and whether I would recommend living off-base. When I got to Pensacola, I stayed in the Bachelor Officers Quarters for three days. I was fortunate in that I knew most of the people I was with from basic school, and already had a group of friends to work with. So right away a group of us moved out to Pensacola Beach. That was in January. And if you can get to draw BOQ, you can actually make some money on the deal. If you get about four guys and can find a house that doesn't rent for too much, you can make out quite well. Of course, that depends somewhat on the condition of the house and what you're willing to put up with. Living conditions out there in most of those beach houses leave something to be desired, but we had a good time and were able to save some money.

Later on, as we were shifted from one field to another, we lost that BAQ money, and ended up having to pay more to live together off-base. But we did it anyway because we were having a good time. Of course, most of us were bachelors and money wasn't a real problem for us. I personally am a great admirer of beach life. Of course, it's a bit difficult to study on the beach because people are dropping in all the time. But what I recommend, Dan, is that you check into the BAQ until you have time to scope out the situation, make some friends and find some people you might want to live with. Then when you find some people who fit that description, you all can check around for a decent place to rent.

Now with regard to training, to answer your question, the Navy and Marine people take their training together. We started out in pre-flight together and the Marines and Navy were all mixed up. When I went through, there were about 25 Marines per class. After pre-flight and survival training, we went out to Saufley Field.

Now there some of the Marines were segregated. There is a flight out there, Flight 18 that is all Marine instructors and Marine students, although there were more Marines than Flight 18 could handle, so some of them were farmed out to Navy flights. But to the extent possible, they try to segregate Marines there.

At Whiting Field there are two Squadrons, VT-2 and VT-3. I went through VT-3 which I think is the better Squadron because they push you a little harder and you get out a little faster. I believe the more you fly, the better pilot you become. If there is a lag between hops at that stage of your education, you can get a little rusty, and it becomes harder to pick it all back up later. I think that helps make you a better pilot.

Then from Whiting we transferred to Sherman Field at VT-6 for advanced instruments and once again the instructors were mixed up, Navy and Marine, and so were the students. From Sherman we went on to Ellison for the helicopter phase and there again it was mixed training. We had Marine and Navy instructors and other Marine and Navy students. In fact, the day I was designated, there were only two Marines designated. All the rest were Navy personnel. So that answers your question. Yes, the Navy and Marine Corps are trained side by side throughout all phases of the program.

Now with regard to shipboard operations, normally when a Squadron is assigned to a carrier, we go aboard one of the LPHs, something like the Iwo Jima, the Tripoli and the Boxer. We're going to be assigned to the Tripoli when we go out initially, and from there we'll go to the Valley Forge.

There will be a battalion of Marines aboard, the Special Landing Force. Our primary function is to provide air support for them. With the SLF, you'll cruise up and down the coast of Vietnam and then, from time to time, land a battalion for various ground operations. Most of the SLF ground operations take place on the coastal plain. That's mainly where we've been working here out of Da Nang.

The Navy doesn't fly support for any of these missions. The Navy does have some CH-46s, which is the same type of aircraft we're flying, but the only thing they're used for is vertical replenishment of the ships. They'll run back and forth from some of these supply boats or whatever you guys in the Navy call your ships, I don't know.

There is always some miscommunication between the Marines and Navy. I am sure we'll have some clashes when I go back aboard the ship and begin talking about the front, upstairs and downstairs, right and left and all that. But anyway, they do fly vertical replenishment and we rarely see them, although I do understand the Navy has some river patrol groups that are working down in the delta regions in the southern portion of the country. These are gunbirds that correspond pretty much to the gunbirds we use here with our helicopter operations.

But I would say that overall the Navy does very little flying in support of the ground troops, and consequently do very few MedEvacs. Now when the

SLF is ashore, we will provide the MedEvac capability for all the casualties the SLF might take. In fact, on the Tripoli, the Navy has only one bird, and that's a tired UH-34 called the Green Bug that's used primarily for admin runs ashore.

You asked about Marine Fixed Wing. We have two bases from which our Fixed Wing operate. Now one is at Chu Lai and the other here in Da Nang. I think there are some Marine pilots operating off the carriers on these strikes that have been going into the north. But primarily the Marines Fixed Wing are used in direct support of ground troops.

Right now the Marine Corps is primarily using, as far as Fixed Wing goes, the A-4s, F-4s and A-6s. The A-6s are used primarily on bombing raids into the north. They're not an attack aircraft so they're really not all that useful in close air support. We've been using them for what are called TPQ bombing missions. I don't know what TPQ stands for, but these bombing missions are completely ground controlled. The pilots take the airplanes off and then control is turned over to a ground controlling agency that sends the aircraft to the dropping point, makes the drop and sends the aircraft in an escape route out of the area and brings it back. Then the pilots take over again to land it. There hasn't been extensive use of the A-6.

Also, the A-6s have had trouble keeping their birds in the air. It's a very sophisticated aircraft and requires many hours of ground maintenance. So they don't have many up and they haven't been used all that much.

As for helicopters, the Marines use basically four types. The Huey gunship which has been configured to carry two rocket pods and six M-60 machine guns. Another version, the Huey Slicks, are used for passenger runs, mainly VIP transportation. There's the UH-34, a tried and true bird, an old but reliable aircraft, used mainly for MedEvacs and also VIP runs. Then they're using the CH-53, the heavy hauler. They don't use it as much as they'd like because they're having trouble keeping it up in the air. Here again it is a problem of a very sophisticated airplane that requires a lot of ground maintenance.

Then there's the CH-46 which is the bird I'm now flying. It is pretty much the workhorse. We do a lot of re-supply work and MedEvacs, and a few VIP and admin runs. Well, not so many admin runs. We do a lot of troop lifts. In fact, tomorrow we have a 12-plane strike going out to look for g**ner rockets. It should be an interesting day, but unfortunately I won't be on it. I've got off duty tomorrow. Also, the 46s are used for recon when we take troops out into the boonies, drop them off, and go back for them three or four days later.

Now, with regard to your question about using Fixed Wing air cover for rescue work, in reference to the film you mentioned, I'm going to call on the expertise of my roommate here, Tom W. Tom, would you please explain the difference in the Air Force's employment of helicopters and that of the Marine Corps?

"How ya going there, Dan? Ace just aroused me from a sound sleep. He says you want to know about how the Air Force and Marines differ in their use of helicopters in Vietnam. Well, you have to remember that the Marine Corps is primarily a rapid-reaction type organization. We use the helicopter extensively nowadays because of its mobility in vertical assault. The Army claims to have developed this, but that isn't true. They have a tremendous system, though, because they have the equipment and the personnel. We tend to think the Marines have the best pilots in the world, and we have a lot of esprit de corps.

However, the Air Force flies helicopters for an entirely different mission. They have several Squadrons of Jolly Green Giants over here, the H-3s, which are used solely for rescue missions of downed helicopter pilots. That is a tremendous airplane, and those guys really hang their butts out the window. They do an outstanding job, but they don't have the type of helicopters used for re-supply like we do. When we fly, we're considered as artillery would be or as any other supporting weapon would be. We're just at the mercy of the grunts. We're pretty busy for the most part just in re-supply missions. I suppose that is our primary goal here. When we operate with the grunts, we use just about every bit of armor and power we have, beginning with the Fixed Wing which have their good and bad points as far as helicopter coverage is concerned.

Let me explain that we do use here for our close air support mission the A-4 and F-4s. Now the A-6s are tactical bombing aircraft and we don't use them for close air support. They wouldn't do us much good. But we do use them for zone preps.

The way the system works is like this. The grunts come up with a tactical operational plan for a field movement. They'll want to go into an area and want us to lift them in for mobility or surprise. Actually, it's not much of a surprise when 12 46s go anywhere, but it's a good idea on paper. We've come to realize that the guys who do the planning over here sometimes have their heads pretty far up the old wazzoo.

But anyway, we'll use the Fixed Wing for initial zone preps and for any known enemy positions in the area of the landing zone, either the A-4s or F-4s, whatever is available. It's hard to say which one does a better job for us. The A-4s seem a little more accurate, but the F-4s carry more. They go ahead and prep the zone, the surrounding ridge lines and predominant terrain around the landing zone. They prep the actual zone last of all to set off any booby traps or mines that might be planted there. As soon as the Fixed Wing roll off cold after their last run, we usually get two or three runs. They drop their bombs first, and then their rockets. Then our Huey Echoes fly in. The Army has the newer H models of the Huey. We have the older Echoes. They carry two rocket pods and four twin-mounted M-60s, and also a chin-mounted M-60 that doesn't work too well. Anyway, their sole mission is to support the 46s on their approach into the LZ and suppress any fire we take. So what will happen then is the Fixed Wing will roll off, the Hueys will roll in and mark the zone for us with smoke. From the smoke we determine both the zone and the wind direction, the latter being real important when you're heavily loaded and want to keep all the wave off capability you have in case something comes up. Anyway, the Hueys mark with smoke and make a low pass to see if they draw fire. If they draw fire, they'll figure the clock direction and either call in more Fixed Wing or try to suppress it themselves.

As soon as they're satisfied, we go in with the troops and set down in the landing zone. We're in constant communication with both the Hueys and the Fixed Wing. As we sit in the zone, if we start taking fire, it's up to us whether or not we want to remain there or keep the troops aboard and lift out. It's pretty touchy, sometimes. You'll be sitting in the LZ, you start to let the troops out and you start taking fire. You have to decide on the spot whether to take off and leave them there, or to stay and risk getting hit by a rocket or mortar as you sit there.

But it works out well most of the time. If we start taking fire, we call clock direction to the Hueys and they suppress the fire as much as they can. Then we lift up out of the LZ and hope we don't take any hits. Meanwhile, the Hueys stay in the area and we hang around too in case we need to go back in and pull somebody out. But by this time the Fixed Wing have run out of gas and taken off. But the Hueys stay around and cover the guys on the ground, make sure they're okay, and that they're moving out from the LZ. The troops don't want to hang around the LZ very long, of course, because Charlie will know where it is from all the air movements. So they need to move out pretty quickly unless it's a good-sized unit.

Well, that pretty much covers the operation over here as far as a strike or an insert is concerned. The Ace covered a few other types of operations that we fly, the MedEvacs and the night stand-bys and the re-supply missions.

Probably the most interesting are the insert operations. They're also the hairiest because that's the time when you don't know what's on the ground and you have no idea what to expect. You have to rely on your own ability and that of your crew, as well as the gun coverage which, I'm glad to say, is often outstanding. It's a tremendous feeling to know that you're working with a lot of professionals over here. The guys take a great deal of pride not only in their flying, but also in their radio procedures and their general ability to handle any situation they or the grunts get into. I don't mean to wave the flag because we make a lot of mistakes over here, and we do a lot of things we shouldn't do. We make lots of errors but we try, like most people, to learn from our mistakes. Still, it's extremely rewarding and satisfying flying over here because you know that a lot of people are depending on you, and often their very lives depend on you.

I'll turn it back to Ace now, having spent all my knowledge of the situation here, and let him continue. We appreciate your interest, and will be interested to hear how you are doing. Stay in touch."

Well, thank you Tom. Now, you asked also I believe about duty stations when we get back. Well, there's no such thing in the Marine Corps as a good deal. In fact, nine times out of 10 you will take it right in the ear. We are given an opportunity to express a desire in terms of our next duty station. We can request the east coast, or the west coast or Pensacola. A lot of guys want to go back to Pensacola as instructors and I don't blame them. It's a pretty good deal. In fact, I put that down as one of my choices.

If you go to the east coast, there is a chance that while there you will go on a cruise, but it isn't guaranteed. The way it normally operates is if you are a Marine, they have you by the balls and they know it. There isn't much you can do. You will spend two years in the States and then come back to Vietnam for another tour of duty.

Helicopter pilots are in a critical MOS and so they're required sometimes to come back over here sooner than are other people of less critical MOS. Well, I hope we have answered all of your questions. Tom and I have enjoyed the opportunity to sit down and do a little BSing. In the future, if you have any more questions, feel free to address them to us. We'd be more than happy to whip off an answer to you. And who knows? We may just see you at Pensacola.

Chapter 14

The Packrat at Christmas

Well, today is still the, well, in 10 more minutes it will be the 30th of November in the year of our Lord nineteen hundred and sixty-eight. I was just playing back the tape with your message to young Tom, and just as you got to the part where you said my middle name was Packrat, Tom popped up out of his bed and began to look around the room. His only comment was, well, no shit. And I thought maybe Tom would like to tell you some things about my collecting that he's noticed since we've been living together.

"Hi Mrs. Ace. Right next to us we have this footlocker and Ace's rack is over there on the other side of the room and mine is right here. We're close enough to hold hands, but we don't of course, being virile Marines and all. Anyway, on this footlocker is about 15 empty cigarette packs, a lot of empty gum wrappers, a few cigarette butts, a broken alarm clock, three pairs of socks, a couple of broken pens, some dog tags, a gigantic J&B box full of trash paper which Ace is saving for some unknown reason, a number 10 can full of odds and ends, plus several week-old newspapers. Now let's see what's on the floor. We have some skivvies, a magazine on G.I. combat, old covers, a shoebox, a broken mirror that Ace uses to trim his mustache with, a few old tape cases, some tin foil with what looks like cookies inside it, a flashlight, a couple more magnetic tape decks, or tape reels I should say, a few midnight readers, a broken shoelace, some plastic knives and forks, what looks like a broken attaché case, a couple of foreign coins, origin unknown, and various other sundry items lying here and on the floor.

Around Ace's rack, yeah, there's a couple pairs of boots, about 65 cardboard boxes and other stuff. Ace is truly a joy to live with because not only does he have that effervescent personality, but he collects trash and leaves it all over the room. It's impossible to move in and out of this place with all this goddamn trash you have all over here.

And then there's the table. You know, we really appreciate the concern from the home front. However, the problem is, right now we've got about six fruitcakes and twelve one-pound cans of stale Christmas cookies lying around among various containers of fig newton's, Hershey wrappers, empty Coke cans, et cetera. And, we haven't even begun receiving the Christmas rush yet.

Bob has just arrived and is gazing over Ace's collection of trash here. This J&B box really gets me. Here's the problem, I got some Christmas packages a few days ago and Ace wants to start wrapping some presents he's going to send off. So he grabbed up all the wrapping I tore off my presents and stuffed it into this box. And now I suppose he's going to glue it all together and make some sort of wrapping out of it."

"Mrs. Mother Ace, this is Captain Bob. You don't know this, but you almost provided us with the solution to our problem with Ace. It seems he was telling me one time that other people have run into this same problem before many times in his youth. And the way this one particular person solved it was to throw all the goods lying all over the bed and floor and dresser and what have you right out the window of the house so that all these various goods would be there to greet said culprit when he came home. And he would finally have to put things away.

Well, Tom and I got the idea a couple of nights ago when Ace left to go somewhere. Tom and I tossed all these goods in the aforesaid manner, and we couldn't help noticing that the room was two-thirds empty. It was raining, too, which we thought was a perfect touch. So, we put all this stuff outside on the wooden catwalk leading to our hootch. Well, about 20 minutes later Ace returned, came right into the hootch, promptly sat down on his bed and began reading again, not noticing a single thing. So what we want to know is if in your next tape you might share with us the rest of the story as regards whether or not what you did proved fruitful or not, because it didn't work for us. Truth to tell, Tom and I had to go out and bring the stuff back in. Well, happy Thanksgiving and Merry Christmas to all, I'm signing off."

I'm back and I just want to make one more comment before I sign off for the night. I was telling young Tom and Bob a few minutes ago that you had once told me that the only kind of woman I could ever live with would be a slob, one who lived in the manner to which I am accustomed. Tom would like to amend that. He said the only kind of woman I could ever live with would be a cleaning lady. I thought you'd be interested in hearing that. Well, that's all for now.

December 4, 1968

Hi folks. Guess what? I won my second Purple Heart today. I got shot down again. Well, not badly shot down. But I had a bird shot up so badly I had to leave it in the field until we could get a maintenance team out to recover it. Took four hits in the bird in an area that had been swept just two days ago for bad guys. But they didn't get them all.

My gunner was wounded. He caught some shrapnel where a round burst shot through the floor. One round sent pieces into my stick. It came through the front of the bird, hit my stick, the top of the handle of it, shattered it, came up and hit on the inside of my armored seat right behind my head, and then ricocheted off through the top of the bird. It was rather exciting. But I just wanted you to know, I'm okay. And, well, the thing I got my Purple Heart for I'm ashamed to tell you. It looked like I popped a pimple on my chin. It only drew a little bead of blood, but that's all it takes to get a Purple

Heart, and it's good enough for me. Some guys have to get their legs blown away to get a Purple Heart.

Oh, well, I don't want to dwell on that. It wasn't as bad as it may sound. Just a splinter, really. A splinter and a drop of blood. I've had worse happen to me there at the house, many times. Well, I'll talk some more later.

December 10, 1968

I got your box the other day, the one filled with the super skates and other things like the Christmas tree and all that stuff. I sure appreciate it. I also got a box from Betty and Grandma. The Smuckers jelly packets were a good idea but they didn't survive the transit very well. They were all smashed and jelly was all over the inside of the box. Other than that, it all worked out pretty well. I've discovered there's a guy in our Squadron who was born and raised in Rising Sun, Maryland. His name is Norman K. So I thought I'd write Grandma and see if she knew his family or anything. It would give me an excuse to write her. I've only sent her one letter since I got here. Maybe one letter.

Well, I've managed to get a little bit of slack for myself here. That little wound I told you about didn't amount to much, but a few days later I stepped in a hole and pulled some ligaments in my leg. So I've been grounded for a while. I've been in my rack for three days which has been kind of nice because I needed some rest. I'll be grounded for another four or five days before they let me fly again; all in all not so bad a deal.

Well, we're getting ready for Christmas, a Squadron Christmas party to be on the 22nd. We're going to have a cocktail hour and then go to the mess area and have something like a mess night. We're going to have the stewards serve the meal by candlelight. And we're going to try to round up all the nurses that we can so we'll have some feminine companionship during the evening. I'm on the entertainment committee. I don't know exactly what we're going to do, but we'll probably just have everyone sing Christmas carols, you know, a simple thing that everyone can enjoy. All in all, I think it will work out pretty well. We had a mess night once before and, although it was a stag night, it turned out extremely well. Everyone had a good time, and I'm sure this one will be just as good. Even if we didn't have any girls come in, I'm sure we'd all enjoy it.

Got another tape from Gene C. the other day. He's getting along very well. He's back in Annapolis, teaching television production at a local community college, and doing a radio show in the mornings. Apparently he's got a pretty good deal going for himself. He seems to like it up there and I don't blame him. I like that area pretty well myself. But that's all the news I have from him.

I got a gift from Davy Stephens and his wife and if you get a chance you might tell them I received it and will send them a thank-you note first chance I get. I have been writing some letters, actually, believe it or not, but I haven't been getting very many. I went there for about 12 days once without getting any mail. But I don't really mind, you know. There really aren't all that many people at home I want to hear from.

Well, we're getting ready to go on board the ship, and I think I'm going to enjoy it. It should be a pretty good deal. Which brings up another question. Did those clothes I sent you from Hong Kong ever arrive? Tom's did. He had some trouble with customs though because they lost his invoice. But he finally got them through and had them mailed on to his wife. I am curious to know if mine ever got there.

I still don't have any idea what I will be doing when I get back to the States, like where I'll be stationed. It will probably be on the east coast. I've requested Pensacola, New River and Quantico. Quantico would be my first choice; I'd really like to get in there. And I guess New River is my second choice because of the availability for cross-country aircraft. I suspect that is where I'll probably end up, and I'll probably get some cruises into the Caribbean out of there.

Haven't heard from my brother in quite a while now. I trust he's alive and well and is managing to keep his head above water. I sure hope so. If you get a chance, tell him I definitely recommend he not fly helicopters in Vietnam. It's not a good deal at all, period. So I would hate to see him get suckered into it too.

So, that's it for now. I'll talk to you later.

December 17, 1968

Had a pretty easy day today, and did no flying at all. I spent most of the day in the rack, one of my favorite places. I did some reading and, as per usual, some deep philosophical contemplation. I do a good bit of deep philosophical contemplation.

Yesterday, I went out and did some flying. We are now supporting a new operation that's just starting out to the west of An Hoa southwest of Da Nang. You'll probably be reading a lot later on this winter about An Hoa because we're gonna be doing a lot of work out there, Operation Yankee Three, or something like that. I'm not sure what the name is but it's a big winter offensive back in the mountains.

Now, we're not actually doing that much work in this offensive because we're flying the Alpha models which aren't really suited to mountain work. We're underpowered for that sort of thing. I guess the biggest load I can carry up on these 1,000 foot ridges is 1,800 pounds. It's kind of interesting stuff to do. In this one zone I had to let myself down between the trees and I could only hover my main mounts. My nose was hanging over a thousand foot cliff. Well, six hundred feet anyway, this cliff. I could look down between my feet and see the ground. But my main mounts on the back wheels were resting on this little saddle back thing, and the blades only cleared the trees by about a foot or two. It was pretty tight, and I had to go in there twice. Once I landed to let some people off, and then I had to go pick up external retrograde gear and take it back to An Hoa.

The other zone we went into was a bomb crater on the top of a ridge, but we didn't have to land there. All we did was come in and drop externals. That was pretty interesting.

Also, the other day I monitored an interesting conversation. It would have been nice if I'd had my tape recorder hooked up to the radios. It was gruesome to the uninitiated, but still kind of interesting.

As the conversation began, I could hear some people talking about how the evening before they had spotted a bunker complex and that they had come back this morning to check it out. They saw this guy out there, this

Vietnamese guy with a hoe. He was working in the field. Well, they watched him a while and then I heard one of them say the guy was shaking so hard that he can hardly handle the hoe.

So then I heard one of them say he was going to look around the area some, keep an eye on that guy, and see what he did. Well, after a while they started to leave. But then they came back on and one of them commented that the guy with the hoe had put it down and was walking towards the river. Well, they kept watching him and directly he got into a boat and headed across the river. So they called into the clearing agency, controlling agency, or whatever, and they asked for permission to fire to keep this guy from crossing the river. You see, the opposite of the river is held by the enemy, and friendlies aren't supposed to be going over there.

So this clearing agency or whatever it is was slow in authorizing these people to fire and they kept saying to hurry up, that the guy was getting away. "Bravo," they said, "He's going to get away if you don't let us shoot. Can we blow him away? Can we blow him away?" So finally Bravo comes back and says, "Roger, you're clear to fire." And the guy comes back on and says, "You mean we can blow him away? Is that what you said?" And then there was a pause. And then this strange voice broke in, I think it was probably someone else flying around who had been listening in, and he just said, "kill him." That was all he said, "kill him." And then the clearing agency came back on and said, "yeah, if you want to just go ahead and blow him away." So they blew him away. They started shooting at him and directly all they could see was his hat floating in the water. So there's a delay and after a while one of them says, "Hey, we got him. He's on the bank clinging to some weeds. We're going to watch and see if he moves, and if he moves we're going to hit him again."

Well, at that point we had to switch channels so I never did get to hear the end of the drama. But it made me feel kind of strange sitting up there listening to those people calmly discuss this guy's fate, you know. I mean it was almost the kind of conversation that you hear between hunters who have spotted a rabbit or some kind of varmint they intend to shoot. And maybe they want to get permission from the farmer before they shoot an animal on his land or something. That's what it reminded me of. And then very casually, matter-of-factly, they just decided the guy was going to die. And he did.

Well, this was yesterday, I suppose, when it happened, the 16th. Also, I made a stop-in out at the German hospital at An Hoa to talk to the nurse out there, and I now have a date for the Christmas party next Sunday. I'll drop in, pick her up, fly her back to Marble for the party, and then take her back to An Hoa the next day, which will be Monday.

I think I have mentioned this before, me trying to get her into some of the functions we have here. So far I haven't had any success because of conflicts with her schedule. But when I dropped in this time, she seemed quite cordial and said she would be glad to come, and also that she would bring a friend with her. It should be a good show. We will start with a cocktail hour and then have dinner. It's steak night at the club so we'll have steaks. We've hired a dance band, too. So we'll have a good time, I'm sure. Probably finish off the night singing Christmas carols or something.

You remember those skates you sent to Tom? Well, those will be part of the Christmas program. We're going to give them to the biggest skate in the whole Squadron. He doesn't know it yet, but it should be good for a laugh. I think everybody will really enjoy it.

So, things have been rather quiet here of late. We've begun finalizing our plans for going onto the boat. Tonight the billeting officer was in our room making plans for who will live together. We helped him out some, as you can imagine. Tom and Bob and I will remain together. We will consolidate with some of the guys next door to us in the hootch, and we'll try to have all the operation duty officers together. We'll have a choice room right next to the ready room. It will be an eight-man room with only six men, so we'll have some extra storage space. That will be a Godsend because we have accumulated a lot of trash. You know how I am about that.

So I'm really looking forward to getting out there and seeing the boat. It's going to require some adjustment because I have a tendency to get seasick. I was out there for about three hours one night and was woozy by the time I got back on the airplane. I was really glad to get airborne again, back in my own element, you know. But it should be a good experience living aboard ship.

Now I don't know after we go back onboard whether we will be there for six months or what. I hope we will be there six months. We're going to be in an all-Alpha Squadron and the boat is equipped to handle only Alpha Squadrons. So there will be two Alpha Squadrons and the rest will be Delta Squadrons. There will be two Squadrons here at Marble, two at Quảng Tri, and two at Phu Bai. Anyway, I'm hoping we stay on the boat for the full six months before I ship out, because you don't get into so many shit sandwiches on the boat. At least not as many as you get into flying out of this place.

Also, since the Alphas are not really equipped to work in the mountains up there west of An Hoa where we're going to be working, we won't be involved in that much of the operation. It's not that I'm trying to pu**y out on this, but I've had my fill of this hairy combat flying. I am more than ready to finish out my tour with nice admin runs. I don't need any more medals.

Well, actually I don't have any medals. My roommate Tom was written up for a Silver Star, and it looks like he may be getting that. The least he will get will be the Distinguished Flying Cross, and he deserves it. He has done some truly outstanding flying. I admire him; he has a lot of balls and he's a good pilot. But I don't really need any of that stuff. I'd be just as happy not to get it. I've got my two Purple Hearts and that's enough.

Well, what else can I tell you? Oh, we cleaned up the room today. Actually, I should say young Bob cleaned up the room and decorated it for Christmas. He didn't even bother to ask my help because he knew it would be futile. He did a real good job, I must say I lay on my rack and watched him work. From time to time, I offered him moral support so I feel like I did my part. Now the key is to keep the room looking this nice for the next 12 days. When we get on board ship, a lot of our stuff will have to stay packed, so we shouldn't have so much problem staying squared away there.

Tomorrow will be another slack day for me because I'll be working as the ODO. We haven't been having very good availability in the Squadron. Today, we only had three airplanes up out of 15. We're short of qualified people as well as planes. For a while there, we were averaging almost 100 percent readiness with 22 airplanes. We were carrying pretty much the workload for the entire group. Even after some of the other Squadrons came here, we were still carrying high availability. Normally, we carried more than 50 percent of the workload. That was a great source of pride for our people; we would look at all the other people and say, Yeah, you pu**ies. But we seem to be losing our edge. Well, I'll close out for now.

December 20, 1968

Well, I have a small amount of tape to finish up, so I'll move quickly. Things have been really hopping around here the last couple of days. I understand the papers back home say about 7,000 g**ners are poised to hit the Da Nang area. I can't say, we haven't seen any of them, although parts of the area out there have gotten hot lately. But we haven't had too much trouble. We've been flying mostly MedEvac missions. I spent this afternoon flying MedEvac. I had two birds, kind of an interesting mission. It's probably the most rewarding flying that we do over here because you can actually see that you are, you know, doing some good. Like today, I went into one zone and picked up a Vietnamese boy who was in deep shit. He was really hurting. They had done a tracheotomy on him, had a tube stuck down in his throat, and they had it connected to some kind of hand pump. They were pumping him up as we took him to the hospital. He was still alive when we got there. The kid couldn't have been more than 8 or 10 years old, he was really hurting. He was with an old Vietnamese guy who must have been his father, but could have been his grandfather.

Mostly we picked up Vietnamese civilians today. We went down into one zone where there were 11 who had been hurt by a booby trap. Two of us went down and split the loads. It was kind of fun.

The other night, just after I finished talking to you, the area just south of the mountain here got the shit mortared out of it, although none of the mortars hit the Marble Mountain complex. They were just dropping them left and right down there. We've been waiting for them to hit Marble Mountain, but so far they haven't done it. In fact, we've started bringing our pistols back to our rooms rather than leaving them down in the Squadron, as we had been doing lately. As a matter of fact, I went out and got an M-16 today, or yesterday, and brought it down to my hootch along with enough ammunition to hold off a battalion.

The only bad thing about it is that if we ever got overrun, we would all be shooting each other left and right. You've got a bunch of aviators scattered all over the base and everybody carries a loaded gun. You know it would be bad. So I think if we ever get hit, I'm going to stay in my hootch and wait until it's over before I venture to stick my neck outside. I'd be liable to get it shot off by one of our own people.

Speaking of that sort of thing, I picked up one guy today that was in real bad shape, and we're not real sure he lived all the way to the hospital. He had a gunshot wound in the abdomen, and it was an accidental discharge by one of his friends. It just goes to show you there are guys over here dying from causes other than direct enemy action. You really hate to see that sort of thing happen.

Well, tomorrow is Saturday, and the day after that is Sunday. That's the day I get to forget about the war for a while. I'm going out to An Hoa to pick up my German honey and drag her back here to Marble Mountain for our Christmas party Sunday night. Everybody's looking forward to that, and I'm one of them. I think it should be a good time.

I don't know if I told you, but I got the socks and skivvies in the package you sent to me. I really appreciate that. They've come in handy. I've worn one pair of socks since I got them, and several pairs of skivvies. If I take my time and am careful, I think I can make the socks last until I come home without ever having to wash them.

Well, that's all for now. Later.

Chapter 15
Aboard Ship

Well, folks here I am aboard ship and, oh shit, this tub just keeps going back and forth, arrgggghhhhhh. You wouldn't believe it. This thing has a round bottom on it and it just rolls back and forth, all the time, day and night, never a break.

Well, I've been getting a lot of experience landing helicopters on carriers, and I'm getting pretty good at it. Since we've settled into a routine, we don't do that much flying, actually. We send about four birds a day into augment the wing ashore. When they get in, they're doing basically the same thing we did when we were at Marble Mountain. We have two birds that work for the ship's admin, like going into Da Nang to get the mail, and running people from ship to ship. That's pretty interesting. But other than that, we get quite a bit of slack, a lot of time without a whole lot to do.

Today, for example, I didn't do a thing. I'll tell you how bad it is. We had an AOM tonight, an All Officers Meeting, where they suggested people should at least try to get into the ready room twice a day so as to read any pertinent information on the blackboard. I did get up early today and fought my way down to the PX and to the post office which is right down in the innards of the ship.

Well, I just don't like it here. The shipboard life is decent and the conditions are nice, if only the damn thing didn't rock all the time. We eat in a carpeted wardroom with nice table and a napkin holder with our name on it. We keep our own napkin because we use linen napkins and they change them about twice a week. The evening meal is always nice because you are served, beginning with soup and salad, then the main course and dessert. Then they come around with coffee. I can say the food is much better than what we were accustomed to at Marble, which was rather tasteless. It was the usual institutional food, but the food here is seasoned and has taste.

The other meals are served cafeteria style with the exception of breakfast, which leaves the noon meal to be served cafeteria style. I figured that out even as I said it.

Tonight we had an AOM and the skipper got up to say a few words. When he was done, he said, "Well, are there any complaints about the ship?" So I stood up and said, Yessir, the damn thing won't stop going back and forth, up and down. Everybody kind of got a laugh out of that. I seem to be the only one really affected that much by the movement, but I tell you it really gets to me, this up and down, back and forth bullshit. Well, I guess I said that already.

I had a pretty nice Christmas. I spent Christmas day flying, got almost 11 hours in the saddle. Believe me that is a lot of time to be flying. I spent the morning hauling Christmas mail to troops in the field. Then in the afternoon I hauled the chaplain around. Since they were all considered to be non-secured zones, I had to orbit overhead while the Chaplain gave his Christmas services. So I had a good bit of flying but it was rewarding to see how appreciative the troops in the field were for what we were doing.

Also, our Christmas party worked out pretty well. I got a date with the German girl, as I told you before, the lab technician from the hospital at An Hoa. I enjoyed that immensely, and had a really good time. She is a decent girl with a lot of class, and I have seen her several times since then. In fact, the

Germans got a big kick out of it because only an American would think about dropping in to pick up his date in a helicopter, which is what I did. I used just a regular helicopter to drive out there and bring her back to Marble. The next morning we took her back on a regular flight, but the flight out to get her was only to bring her in, courtesy of the Marine Corps, the U.S. taxpayer and of course, the old Ace.

She brought a friend with her who spoke little English. That proved no great problem because we have a man here in the Squadron who was born and raised in Germany and speaks the language fluently. He was able to entertain her friend. It was probably fortunate that he is already married because her friend was rather ugly and if he wasn't married he would have probably wandered off to scout out some of the other ladies.

Before I left Marble, I drove out to tell her I was leaving. Every time I went out there it was kind of difficult because I had to stay with my bird. It's hard to talk beside a turning helicopter. I did go out one day to take her a Christmas present. It was a bottle of after-dinner liquor she and her friend had expressed a preference for.

So I had a good Christmas. I got several gifts from different people and was very appreciative of their efforts. We saved all the wrapped presents until Bob got back from Hawaii from his visit with his wife. As soon as he got it, we had a little party of our own, opened all the gifts we had gotten. It was kind of nice.

Of course, we spent New Year's aboard the ship. It was just another day like any other day. However, I have just been looking at the calendar and I see I'm getting short. I'm on the downhill side now. I've only got something like 27 weeks to go which is not all that long when you get down to the nitty-gritty of it. I'll be getting home sooner than you think.

I'm going to Sydney here in a couple of days. Well, in a moment we're going up to the ward room to watch a movie. Before I go, I wanted to tell you that I was down there the other night playing bingo, a dollar a card. A lot of laughs. I thought you'd appreciate knowing that warriors out on the open sea in the combat zone while away their evenings playing bingo. Well, I've got to make a head call and get on down to the show. So I'll check you later.

January 26, 1968

Hi folks. This is the Ace speaking to you from his stateroom. I use that term loosely, but it is the Ace, the scourge of American motherhood, speaking to you again. It's been quite a while since my last tape; three weeks in fact. I mentioned I had only 27 weeks to go. Well, make that 24 now. So it won't be long, only one hundred and sixty some days, until I'm back in the saddle again.

Well, things have been happening pretty fast and furious around here the last few days. I think I'll start with today and work backwards, just to keep it all straight. I was out flying test hops this morning, and this afternoon I'm on MedEvac stand-by. We have a single bird MedEvac and I'm hoping we don't launch this afternoon.

Yesterday, I went out for an admin run in the morning and ended up getting diverted to an emergency MedEvac. I just happened to be the only available bird in the area. We're right now supporting Operation Bold Mariner down south of Chu Lai, in Quảng Ngãi province. So far the operation hasn't been very exciting, which is all right with me. I am sure the rest of the people around here would echo that sentiment.

But anyway, we haven't had too much enemy contact lately. The Majority of things our people have run into are mines and booby traps. Yesterday, the 1st Platoon of Mike Company of the 3rd Battalion 26 Marines walked into a world of shit. They got into a mine field and before they got out, they had taken casualties, 3 KIAs and 15 WIAs. I got diverted out there for the initial MedEvacs when things first began happening, and I had 15 WIAs on board. That's a hell of a load. They had all their gear and everything. They also had three KIAs out there, and I decided not to take them. I just lifted out the WIAs and took them back to the ship as fast as I could. Then when I got back I found out that three of the people I had on board were very serious and that I had to take them on to the hospital ship that's floating around out here with us, the U.S.S. Sanctuary. Took them over there, dropped them off, and went back to get the three KIAs. I discovered that in the intervening 10 or 15 minutes, they had taken four more WIAs and two more KIAs in the same area.

Fortunately, they were not under enemy fire when I got back out there which would have made things infinitely more difficult. I found out later towards the end of the day that the area in which I had been landing to pick up these MedEvacs was right in the middle of the mine field. In the afternoon, one of the other birds had gone into the same position to pick up some MedEvacs and two guys were carrying one KIA to the aircraft, when one of them stepped on a mine. It killed one of them and wounded the other, making altogether two KIAs and one WIA.

The guys on the ship said that in all the time they've been here they've never seen anything quite like that. They had to set up the hangar deck as sort of a makeshift hospital receiving room because they didn't have room downstairs in the regular ship's medical quarters there. The ship is pretty well equipped to handle these emergency cases, with several operating rooms and teams on surgeons on call, available to work. So yesterday they took the more serious cases down to the operating rooms, and handled the less serious ones right there on the hangar deck.

Things got interesting for me towards the end of the day. I'd just brought some MedEvacs back and they had two that had been treated downstairs, but were very serious and needed to be transferred to the hospital ship where they have better facilities. I had just lifted into a hover on the deck when I got a compressor stall on my number two engine and lost it. Fortunately, I was able to set it right back down on the deck. But had I moved another three feet or so, I would have been in deep shit.

I figure somebody must be looking out for me. Had I lost that engine when I lifted out of the zone with those 15 MedEvacs on board, I'd have been in deep trouble too because even as it was I lost some turns. I don't know if you're familiar with this, but if you lift a very heavy load in a helicopter and you move out of your ground cushion, and don't have sufficient power, your rotors began to slow up. It takes a certain technique known as milking to try to get your turns back to maintain your forward air speed and your altitude. But I drooped down about 92 percent getting those guys out of there. Had I lost an engine then, I would have been hurting. I would have come right out of the sky.

One of my roommates pranged one up the other day. He lost an engine and almost rolled it up in a little ball. Everybody got out of it okay, but the airplane was pretty much a total loss. We've had a rash here in the past week or so of losing number two engines. I lost that one yesterday, and the day before I almost lost one, but managed to hold on long enough to get it on the ground.

This particular time I had 52 people onboard, if you can believe that. We were carrying detainees, Vietnamese they've rounded up in the operating area there on the peninsula. We were carrying them over to a detainee camp where they interrogate them to determine the good ones from the bad ones. I had 52 on board; women, children and old men. You can get a lot of Vietnamese on an aircraft because they don't weigh very much. You're more space-limited than weight-limited with these people.

But I was on my way when all of a sudden I began to lose oil pressure in my number two engine. Well, I immediately began to think about getting it on the ground, and I was able to get it down before the engine quit. But had I lost the engine before I got it down, I think I would have been in deep shit because I don't think that my single engine capability was very much, something akin to a brick.

Other than that things haven't been too exciting around here. We've been flying mostly re-supply to these people out there, hauling ammunition, food and water to them as they move. Originally, there were two battalions in the operation, the Second and Third Battalions of the 26th Marine Regiment. The Second Battalion is gone now, and along with them went the U.S.S. Okinawa, which is another LPH. So now we only have the Tripoli, which I'm aboard, and several other supporting ships from the attack transports. There's also some LSDs. I'm not sure what they are exactly; the Navy has a lot of stuff that's hard to understand.

Well, let's see. While I was gone, oh yeah. I maybe didn't tell you about my trip to Sydney. I was gone for six days. While I was gone, one of my other roommates, Jerry S., pranged one into the side of a mountain. They all got out okay, but the aircraft burnt up into a little ball. They brought what was left of it back in a little basket.

We've been really fortunate in this Squadron in the last seven or eight months in that we haven't lost anybody. We've lost several airplanes and had people wounded, but no one has been killed.

Now I'd like to cover my sojourn down to Sydney. I left here by myself. No one else from the Squadron was going that way at the same time. But when I got over to the R&R center at Da Nang, I ran into another Marine, a First Lieutenant with the First Marine Division, who has been serving up around the DMZ as an artillery officer. Also I met an Army First Lieutenant whose unit was involved in this operation down at Chu Lai. It turned out that he and I had chosen to go on R&R at this particular time for the same basic reason, neither of us wanted anything to do with the initial assault phase of this operation. He only had 50 days of his tour left.

Anyway, we left Da Nang and flew to Sydney. We stopped at Darwin for a while, long enough to get a buzz on upstairs in the bar there, before we flew on to Sydney, getting in there about 8 a.m. local time. We had to go through their customs, which are pretty strict. They took all the cigarettes; I think they were looking for marijuana. We've had some problem with that here in Vietnam, primarily among the enlisted personnel.

Then they began to search the airplane. They let people go one row at a time and held them at the front of the airplane while they searched everything. Then when nothing was found in your seat, you could go on to have your luggage checked. From there, they loaded us on a bus and took us to an R&R center. There we got a brief on what was going on in Sydney, the customs of the people, and things to watch out for in order to stay out of trouble. We also exchanged our money and made hotel reservations.

After that, they had buses take us to our various hotels. I stayed together pretty much with the Marine, whose name was Tim W. The other guy, the Army Lieutenant, was named Ted. We all started out together and the first night we get out to meet some girls. There was a nice bar there in the hotel, one that required a coat and tie, but Australian bars are different from American bars. You have to stand up all the time, even in the nice ones, which I didn't care for very much.

Well, anyway, after a while I spied these two girls and suggested we go talk to them. Tim said to wait until something better came along, but after a while he said, well, okay, let's go. So we struck up a conversation, invited them to dinner and then I got cut out. I'm not sure how that happened, but it ended up with two couples plus the Ace. I must have done something wrong, which would not have been the first time. So after a while, I struck out on my own.

It got interesting. I was in this one place, oh it must have been about 4 a.m., and I was just blasted, you know? And this guy comes up, grabs me by the arm and says, "Are you an American?" So I said I was, and then he demanded to know if I was in the military. I said I was and then he had to know my rank and everything. Well, he started going on and on about how he had to get into the American military and go to Vietnam. It didn't take me long to figure out this character had a real mental problem.

I spent much of the next day recovering. The other guys had set up dates with the girls the next night and they promised to find someone for me. Well, they were supposed to, but communications got fouled up somehow. First, we went to this mixer that the R&R center arranged for us, and while we were there Tim met this other girl and decided he liked her better than the one he had met the night before. And he just up and decided to go off with her and stand the other girl up.

Well, being the gentleman that I am, I decided that wasn't quite right. So I went along with Ted figuring I would fill in for Tim with the other girl. You know, I figured going out with the old Ace has to be better than staying home alone, right? So I meant to break it to this girl gently that Tim had gone off with someone else, and I would be her date. But when we got there we discovered these girls had arranged a date for me. So I ended up with two dates. And it got even more interesting later on when we ran into Tim and his new date and they joined us for the remainder of the evening.

Anyway, we all went to this restaurant where the Army guy, Tod Stoffer, had made a reservation for all of us. By that time, there were nine of us altogether, I forget where the others came in. Anyway, when we got to the restaurant we discovered that some jerk had come in and claimed to be Ted S. and changed the reservation from nine to two. I guess the guy didn't have a reservation of his own so he just looked at the list and claimed a name. I mean, how many Stoffers are likely to show up at a restaurant on any given evening? Well, we let it pass. We eventually got a table and had a truly excellent meal, really good.

Later, we went out and hit some wine bars which apparently are the big thing in Australia these days. They don't sell anything in them but wine. Well, that was fun, but you get a little dizzy after a while.

One day, Ted and I rented an airplane, took an instructor with us, and went up to fly around, check out the landscape. We had sort of parted company with Tim by then, because he wasn't being quite honorable with us. Anyway, I was really surprised to see the real Australia. You have this stereotyped image of flat land and desert, which is, I suppose, most of Australia. But there along the coastal area it's mountains and forests, rivers and lakes, just like back in the States. It's just a belt of green and water along the coast, but the belt is much wider than I had envisioned.

After the second night, the girls started taking us to parties. The one I was supposed to be with the second night, Tim's original date, didn't show up any more. I guess I don't really blame her. It was kind of shoddy treatment she got, despite my gallant effort to compensate by being a good guy. Of course, I'm always a good guy. But anyway, we didn't see her anymore. So I ended up dating another girl, a roommate of Ted's girl, who was a model. These were really classy girls, by the way, real classy.

The third night we went to a party thrown by some Australians where we met some interesting people. And more parties night after night. Just lots of parties. The Australians like to party and they also like Americans. The last night the girls hosted a dinner party in our honor, which was really nice. So all in all, my trip to Sydney was a rousing success. I really enjoyed it. I hated to come back to Vietnam.

Oh yeah, one more thing. I made some purchases in Sydney and have arranged to have everything sent home. The main thing is that I bought a tape deck here that I should have sent home but didn't and now I have to lug the damn thing around everywhere I go. I should have shipped it home in the beginning. I didn't want to make the same mistake with the stuff I bought in Australia.

I bought a nine by twelve foot kangaroo fur rug, and also a kangaroo suede jacket. I also got two kangaroo pelts. They are just the skins, you know, with the tails, just like they came right off the kangaroo. Also, I got two sheep skins. All this stuff should be arriving within about a month.

I also got a family crest made and framed, a crest for the Astle family on parchment. While I was there, I called up some Astles I saw in the telephone book and had a really interesting conversation about our family tree. There is a bunch of Astles in Sydney, believe it or not. They wanted me to come around and visit, but there wasn't time. Sydney wasn't expensive at all, not nearly as expensive as Hong Kong. I took along $700 and only spent $600, which is unusual for me, to come back with money. Which reminds me, I'll keep sending $200 a month home. Then I'm going to start saving out the rest to build a little nest egg over here. I am discovering that a lot of the money I send home is going to pay bills, and a not inconsiderable portion is being loaned out to my little brother. Not that I mind, but it may be a while before my little brother manages to pay me back. Of course, as my roommates say, I'm easy. I won't really care if I never see that money again. Oh, of course, I would like to see it again. Don't you dare tell Dave any of this or I for sure will never get that money back.

When we got back from Sydney into Da Nang, it was dark and the weather was shitty. The ceiling was low, so apparently the pilot elected to shoot a TACAN approach into the landing area. He broke through the clouds and I couldn't help noticing we didn't have much altitude. Also, he was flying a straight path across bad-guy country, for which reason I was sorely concerned. There are just uncountable bad guys in that area. In fact, at night around there I think every farmer digs up his rifle and shoots at airplanes. So I was watching out the window and, sure enough, we were taking fire. And I will tell you the truth, it just scared the hell out of me when I saw those tracer rounds coming up. It was a hell of a welcome, you know, back to the war. But they didn't hit us and we got back into Da Nang in one piece.

Then I had to wait another day to get back to the ship. When I left for Sydney, the ship was in Da Nang at the harbor here. But when I got back, it was 70 miles out to sea. But I managed to get back eventually.

They've been flying me quite a bit since I got back trying to get me caught up hours-wise with the other pilots because I got off to such a slow start this month. But I have about 35 hours now and am no longer the low man on the totem pole.

My roommate, Bob W., just got back yesterday from Hawaii. He'd been there visiting his wife and is still walking on the clouds. I don't blame him. I think I probably would be too.

I'll tell you, it's the most depressing thing in the world to come back to this place after you've been away. This place really blows. I really want to get away from here so bad.

I regret I don't think I will be able to attend Dave's graduation. Getting back to the States is just not in the cards, even if it is to watch a brother graduate from the Air Force Academy, not unless he can pull something out at his end. I would really like to be there, but I'm sure he understands. In Pop's last letter, he mentioned my motorcycle, about getting it fixed up. I didn't really mean for you all to hustle right out and do it. What I really meant was for you to wait until closer to my homecoming and then do it. Also, I would like to get it licensed in Florida again where I bought it. I can't see any good reason why I should have to pay property taxes on it in West Virginia when I haven't hardly been there.

I expect I'll find it hard to readjust to the civilian world. Pilots are treated like prima donnas over here; we kind of run our own show. When you're out there flying around working with these grunts, you tell them what you want and they do it. It's supposed to be the other way around. They tell us what they need and we support them. But we feel if we're going to support them, we've going to do it our way, and this pisses them off a lot.

The other day I went into a zone and because of the existing conditions, the wind and the way I shot my approach, I screwed it up a little bit. I came in a little fast and overshot the zone. Then I put the load down in the wrong place. Christ, it couldn't have been more than 25 feet from the target zone, but it was in a different rice paddy. Then the guys came on the radio about it and they were really pissed off. I told them to forget it, that I'd fly my airplane and let them run their own show.

But it gets to be a habit, you know, telling people how you're going to do it. It might cause some problems when I get home and start telling people how I'm going to do things.

One thing, I think this whole experience has been good for me because it has given me a lot of experience making snap decisions, snap judgments and so far I haven't made any bad ones. I feel good about that. As Pop probably remembers, I used to have a tendency towards being very indecisive and not being able to make up my mind. Working over here you have to make up your mind, usually fast. It becomes a habit after a while. Personally, I think it's a rather good habit. It gets things done.

That reminds me. When you get the bike fixed and registered and all, I want you to get it insured. I think United Services will do that for you. I'm not sure what their policies are on motorcycles, but I've had a good record with them since I got my car insured. I haven't had any claims or any tickets, so it shouldn't be a problem.

You know it's funny, but I made the observation the other day about how similar the life we're leading is to that on a prison trustee farm, like the one there near Barboursville. You know, where they don't actually lock them up and they do work. We're like that. We work and they don't actually lock us up, but we can't leave. The things you take for granted in civilian life, like being able to jump in a car and go where you want to go, are missing here. We really miss that, or at least I do. I like my freedom. One of the things I'm really looking forward to is being able to do what I want when I'm not on the job.

I got a letter from Lieutenant R.T. the other day, and I note with interest he is getting out of the Marine Corps. I was a little surprised to read that, but then again, I'm not really because I guess it just took him a little longer to see what was going on than it took some other people. But I agree with him 100 percent.

The Marine Corps has a problem now retaining junior officers to the point they did a study of it here in Vietnam. We were asked to fill out a questionnaire explaining why we intend to leave the Corps. They think it's the money, but that isn't it. The guys are getting out of the Marine Corps because of the way the Corps is operating. For example, the leadership of the Marine Corps has been a great disappointment to me personally. Some of these Majors, and some above the Majors, have really been poor excuses for leaders. I think that is the main problem. They always preach about how a Marine officer is a leader of men, and he will be called upon to make decisions, and he will be given a certain amount of latitude to handle things as he sees fit. But this is not the way it operates, not at all. It's gotten to the point now that the junior officers are afraid to make decisions on their own because if anything goes wrong, they'll just get shit on because no one will back them up. So they just kick the decisions back up to the Senior Officers, which is what they seem to want anyway.

They don't want their junior officers making decisions on their own. The result is a general refusal among junior officers to accept responsibility.

Now one good thing about being in the Air Wing is when you're flying, there's nobody who can take responsibility for your airplane but you. This is something that must be accepted by the man who signs for the bird. Any decision you make, you know it's up to you. People can say you were wrong but they can't really nail you for it because they weren't there. But in other areas of command, the junior officers simply will not make decisions because the Senior Officers won't let them.

Take scheduling for instance. You know, I was the scheduling officer for a while. But this Major just wouldn't let us use our own initiative in writing the schedules. He wanted to look right over our shoulders and approve or disapprove everything we set up. After a while you start to think, what the hell, why not just let him write it. Why waste your time? In fact, I think that attitude is the reason I got shit-canned from schedules. At the time, I was a bit upset. But I'm glad to be out of there now because that Major is still driving everyone crazy. You'd think he was CO of the Squadron and not just the operations officer. Fortunately, I don't have to deal with him anymore.

In fact, I don't have any job at all, not really. They gave me the nebulous title of the Pilot Ground Training Officer, but we're so busy there's no training to be done. I gave one lecture just before we came on the ship, but I didn't really prepare for it. I just sort of whipped it out of my ass when I was up there in front of the group and it probably showed. But I didn't care. I mean, what can they do to me? Make me a helicopter pilot and send me to Vietnam?

They really don't have much to hold over my head. They can write me up in a bad fitness report, but then what will they do with it, stick it in my eye? If I were planning to make a career of the Marines that would be different. But I don't care. I do my work, and so far as flying goes, I think I do a good job. But some of this other stuff they ask us to put up with is just too much.

"MedEvac inbound, bravo six, man your stations. MedEvac inbound two minutes."

I don't know if you just heard that in the background, but they just got a call MedEvac inbound, bravo six, man your stations. MedEvac inbound two minutes. That's what we hear when they bring a MedEvac onboard here. You might be able to hear the sound of the aircrafts up on the deck. My stateroom is directly under the flight deck. In fact, the ceiling is the bottom side of the flight deck, so we get a lot of noise.

Speaking of MedEvacs, you know, yesterday when I was carrying those people, God, that just tears you up to see that. It's such a waste. I sat there in the zone and watched and they put 15 people on the airplane. Some of them were walking wounded who weren't hurt so bad. But some of those poor guys were just torn all to hell. And you just look at them and wonder what's the fucking use, you know? It really bothers me when I see that kind of stuff. Now 15 more people are dead. For what?

Then last night we learned that one of our Squadrons up north lost a bird. I had a good friend who was HAC on it, a guy I went through military training command with. They were out on a night MedEvac and drove it into a mountain. Killed everyone on board. Sometimes it seems like you're just batting your head against a wall, especially down here where they really haven't accomplished anything. They haven't found much or killed many enemy troops. We've just kind of had our asses handed to us down here. Most of it has come through booby traps and mines.

But anyway, getting back to those MedEvacs, the only thing about them is that you feel like you're doing something worthwhile. At least the guys really need you. I'm taking these guys from out here in the boonies into a hospital. They're going to live because of me, particularly the guys who've been hurt the worst.

But sometimes it just works on you. They had one guy they laid on the floor just behind the cockpit, the bird being so full and all. They laid this guy right beside the openings and he didn't have any face. You just look out the aircraft and you look around when they start loading them in the back, and you see shit nobody should ever have to see.

It doesn't really turn my stomach, but it's revolting all the same. And still it's hard to take your eyes off them, you know. You really feel for those guys. I tell you, there's no way I'd want to be on the ground there. I'll take my chances flying. I don't envy those guys at all. Like I said, they really took it in the ear and the whole first platoon just kind of ceased to exist. I don't know. I just don't know.

Chapter 16
The Wrath of Grapes

Hello family. Today is Wednesday, the 29th of January, here on the ship in the ready room. I just thought I'd go around the ship and pick up some sounds that we hear every day for your enlightenment. (Loud machine noises) There, that was the sound of the weather machine.

We are now in the midst of a game of acey-deucy, quite a popular sport around here. If you listen carefully, you may also hear the tell-tale sound of dice hitting the table. Back at Marble we used to run these games all the time, and being aboard ship hasn't dampened our enthusiasm any.

Now I'm standing up on the flight deck looking out toward Vietnam. I guess we're no more than three miles offshore, maybe less. They've got two destroyers between us and the beach. They've been laying in some gunfire on targets inland. I understand that this morning the ground troops discovered a large tunnel complex, which they are now working over. We've gotten in five KIA on the ship so far today. I'm not sure what the WIA count was.

That sound you hear in the distance is a 46 coming back into the ship. Actually two of them in flight. I'm not sure, but it looks like they might be coming in to pick up some externals. I was supposed to be a part of that flight, to lead the second section, but we got scrubbed at the last minute. They decided they only needed two birds which is why I got this break to walk around the ship talking to you. Now here comes another 46. While that was going over I saw the smoke from an explosion on the beach, but the helicopters blanked out the sound.

Now we're up into the tower overlooking the tower. (Sounds of radio transmissions, telephones ringing, voices talking.) And now it's quite evident from the background noise that I'm back out of the tower. I'm on the same deck level looking down over the flight deck. (Sounds of helicopters) Our mail came in just a few minutes ago, so I may go see if I got anything. Why look! There goes my roommate Bob W. lifting off the deck. Now I'm going to go back down to the flight deck, and you will notice the noise level. (Noise level increases to low roar.)

February 5, 1968

Well, the operation is almost over now. We started retrograding the gear and some of the people today, although we still have two more companies left to bring in. I was supposed to fly SAR, or Search and Rescue this afternoon. But by the time they got my airplane spotted on the deck, the other birds had just about finished. Some of the people didn't have all their shit together so they came back early and I didn't get to go out. I got turned up on the deck but didn't get launched. It's already the 5th and I've yet to fly this month.

I got your tape yesterday and played it over a couple of times. I really enjoyed it; it was a good one. You had some of my favorite people on there. I have a couple of messages I would like you to pass on.

First, tell Mr. Childers he is very perceptive and that about three days ago I put in for transition to C-130s. So he's right. And tell Mr. Avis I haven't forgotten that steak dinner, and I fully intend to pay for it as soon as I get home. However, in the intervening time he should give a great deal of careful thought to what he's going to bet me next.

Mom, you asked about where we're working. We've been working in the Operation Bold Mariner which is taking place a little bit south of Chu Lai. Now as soon as this operation is over, and we should have all of our people out by tomorrow, we'll be heading back north to Da Nang. We will stay there for a while until these people get put back together again. They really got into a big hurt down here when they ran across that mine field, took a lot of casualties. So we'll be up there for a few days, and then our ship, the Tripoli, will be heading for the Philippines.

Now at that point my Squadron is going to split up. Some of us will be going to Phu Bai and some on to the Philippines in the Tripoli. Once they get to the Philippines, they'll have four birds. They'll be used to offload gear onto the Valley Forge which is the next boat we will be working on. Then they come back to Vietnam. Of course, by that time it should be the first of March or close to it. I am kind of hoping I will be with the bunch assigned to go to the Philippines. For one thing, it would get me out of the country during the initial phases of the Tet holiday.

The g**ners have declared a truce, a seven-day truce for Tet, but they talk out their ass a lot. You can't believe anything they say, so I don't think it's going to be all that good.

"Bull sh*t!"

There in the background you may have detected a strange call. Well, that's the call of an enraged Weinberg sitting over there on the other side of the room engaged in a heated argument with somebody. We were just picking up some of the verbal explosions coming from over there.

I wrote a letter to Dave. Told him I had spoken to several people in the Squadron here about maybe getting leave to come to his graduation, but it doesn't look too promising. Of course, it's difficult to explain to officers who are short of pilots that you would like to leave. They just don't like to send people away for things like that. But you never know. We'll have to wait until the time gets a little closer and hope something turns up.

I do have one more thing I'd like to say. I think it might be a good idea for Pop to give up his mechanic work. He says he broke a sparkplug, or a wrench on a sparkplug on my car. I wonder if he forgot which way they turn. Tell him to take it to someone who does that kind of work for a living. You know, they've got to eat too.

The weather the last couple of days has been really shitty. Ceilings have been down to about 250 or 300 feet, and visibility down to about half a mile. We were out flying in that today. I didn't have to launch, fortunately. I didn't really want to have to get out and fly around in that kind of stuff. I'm hoping the weather will break soon. If the old wives' tale about the groundhog is true, then the weather should break, because there is no way the groundhog could have seen his shadow around here on February 2nd. The monsoon season will be over toward the end of this month and it should start clearing up.

It's getting pretty short now. I guess maybe I've got about 156 days, something like that, four months, well, five months. February will be over before long. I guess we're about one-sixth of the way through February already.

I was glad to hear Jim T. has a girlfriend now. That should help him out pretty well. I think he's needed one for a long time, and I hope everything turns out well for him.

Well, it's getting near time for the movie, so I'll close out and talk to you later.

February 10, 1968

Well, Operation Bold Mariner is officially over as of yesterday. We finally retrograded the last of the people off the beach and got them back on the ships. They really didn't have much of a break because today we put them back in the field again, this time up by Da Nang. We put in a bunch this morning and the rest this afternoon. I got a little over six hours in today, almost seven actually. But it was broken up, so it wasn't so bad.

Well, I have some bad news for you. I don't know if you remember Andy H. or not, but he was killed about a week ago in a mid-air collision. Kind of a bad thing. In fact, it hasn't been going all that well for helicopters up here the last couple of weeks. We had the crash Andy was killed in. Then we had a bird shot down not too far from where we went in today. It got shot up by a 50 cal, crashed and burned. Nobody got out of that one. Then we got word that another friend of mine, a guy I met when I first went to Quantico back in '62 crashed in Da Nang harbor. The only one to get out of that was the crew chief. They found him floating in the bay next morning. He had been floating around out there for 10 hours.

It kind of makes you wonder, you know, when you start losing friends like that, particularly when you're starting to get a little short. I'll tell you what. This kind of thing is exciting when you first get into it, maybe even a little fun, but it gets tiresome after a while. I've decided I'd like to get maybe a group job where I could go out and fly a couple of times a month on admin runs, and spend the rest of the time working in a nice, quiet office someplace. But I don't think I'm going to be that lucky.

In a couple of days we'll be offloading at Phu Bai just in time for Tet. Word has it that the Tet Offensive this year is going to start on the 12th. You see, Tet officially begins on the 17th. But last year they began their Tet Offensive on the 17th and it just pissed an awful lot of people off because it was right in the middle of the holiday. So this year they're going to start it on the 12th and end it by the 17th so everyone can get in their celebration. Well, of course no one likes to fight on a holiday. That sort of thing just isn't done.

We got our cruise books the other day. A cruise book is sort of a military yearbook. It covers a Squadron's operations for a year and has pictures and stuff in it. We were in the Squadron for six months of the year and were fortunate enough to get our pictures in it. I didn't get quite as many pictures of my pretty face in it as I would have liked, but I'm already doing better for next year's edition. Every time I see someone with a camera I get in the range and smile as big as I can.

Yesterday, I got the kind of deal that just goes to show war isn't all bad. I came up to Da Nang from the ship and picked up a USO show. They had five girls and three guys. Tom had the other bird. We flew up together and we left early so we could get up there, shut down, pick out the girls we wanted on our airplanes, and bring them back to the ship. We watched the show and afterwards took them back to Da Nang. The trip back was interesting because it was black last night, no horizon at all, not even a moon out, and it was hazy too. I had a little trouble when I first lifted off the ship. You know, you're sitting there with all of those visual references, the lights and other things on the ship. But the minute you lift off it is nothing but black void. There was no way to tell the altitude other than with instruments. It's kind of hard to just go immediately from visual to instruments like that. It takes a few seconds to adjust your control responses to the visual representation that the instruments give you rather than what you see with reference to the ground. It gave me a kind of scare there in the beginning, but I got it under control. We had to fly instruments all the way to Da Nang. We got there and landed, which was also exciting, landing on instruments. It was an unlighted zone.

Then I let my co-pilot fly us back. He did a pretty good job until we got to the ship at which point he damn near put us in the water. He got fixated on the LSE, you know, the guy who stands out there with the lights and directs you into your spot. He started getting too low and I told him. He kept getting lower and the tower mentioned it to him, also. Finally, I had to grab the collective from him and pull it up. The thing that bothered me was I was calling off altitudes and air speeds to him, and his rate of descent, but it just didn't seem to register in his mind. But we got it under control and made it back to the ship. It's easy to see how these accidents can happen when you're not on your toes.

This thing we did today was kind of beautiful because of the way it was handled. We have this Major who is a Squadron executive officer who is in something of a trance most of the time. We call him Teddy Bear because that's what he looks like, a big teddy bear. His last name is Ribbeck. So now the Squadron joke is Major Ribbit, Ribbit.

Anyway, Major Ribbit almost sent us all to our rewards today. We launched from the ship and coordination was a bit poor to begin with. We were supposed to have Huey gunbirds, but we hadn't gotten in contact with them. So Major Ribbit comes up on this fox mic frequency which every g**ner in the world monitors and without even establishing radio contact with these people he begins to transmit in the blind telling them that they were inbound to the zone, gave the coordinates to the zone, the number of aircraft we were bringing in, the approximate time the lift was going to take, the whole works.

Had there been any g**ners out there, they would have really zeroed in on us. He just really compromised it. He's pulled several winners but his day will come. You see, we have this kangaroo court now in the Squadron. Every time somebody screws up like that you can come back, write up a charge sheet on him, and drop it into this box. And then periodically we hold the kangaroo court, and people get fined for various offenses. You can get fined anywhere from a nickel to a quarter. We held one the other night and made $5.45 for the Squadron. We're going to use the money when we get to Phu Bai to throw a party. That's when we get back to where there's a bar again, off this dry ship. But I'm not all that excited about going to Phu Bai, to be honest with you.

I had a pretty good day mail-wise, got seven letters which was nice. It took me a while to get them all read, and now I have a lot of work ahead of me to answer all of them. So I better sign off now so I can get to it.

A few days later....

Hi folks, here' the Ace.

I hadn't talked to you all since earlier in the month, and since I spoke I made the 18,000th landing on the Tripoli. I got a picture and all to prove it. Plus a plaque with my name engraved on it, which I'll probably be sending along to you soon. But we have moved to Phu Bai and tonight is our first night here. And it may not surprise you to learn that Tom and I have been partying some. Tom, would you like to say word?

"No, I'm reading a letter from my fucking wife."

Um, Tom is reading a letter from his wife, so we'll let him alone for a minute. But he'll be ready to talk in just a second. We'll go up here and see old Jerry. This is Jerry S., another roommate. Jerry, would you like to say a few words to the folks at home?

"Hi folks, how are you doing? How are John's folks?"

That's fine, Jerry. Say, do you like Phu Bai?

"Oh, I love Phu Bai."

That's what I thought.

"Ace, do you like Phu Bai?"

No, that's a negative. I'm not even going to talk to you anymore.

"I had a chance to move out of this place tonight, did you know that?"

You had a chance to move out? Why didn't you?

"I didn't want to leave you guys all by yourselves."

Aw, you asshole. Jerry is a good guy. He didn't move out. He's living with us in this, in this, well, absolute squalor. Now Tom, do you wanna talk now that you've finally read that letter?

"I have more fucking letters to read, Ace. I'll be with you in just a minute."

Okay, listen, we'll be with you in just a minute as soon as you read all your letters, okay? Okay. I will talk more in a minute.

Later.

Hi Folks. This is the Ace talking again, and this here guy with me is Tom.

"Ace, you're a good shit."

Listen, this is the second night we've been in Phu Bai, and this is the second night we've been drunk because they don't really need us. Dumb shits. They wouldn't let us go to the Philippine Islands with the ship because they said they wanted to have another three 46 Squadrons at Marble which gives them more airplanes than they need. They only use four to six airplanes out of Phu Bai a day. And now they got another Squadron up here, now they got two Squadrons. Tom, you missed us barfing out there, so we…

"…another barfing session, that's the whole point of the tape."

Well, I had to explain it to them. Now, why do we go out to barf?

"Okay, here's why. You're in a bar, you're drinking sociably, right? You sit there and you drink and you're playing a game like Whales' Tales. Whales' Tales is a good game."

And Whales' Tales. Or Thumper?

"Or Thumper. You're drinking sociably, so you drink pretty fast, right?"

Right.

"Especially when you're dumb and you're losing. Most guys are dumb and they lose the game. So what do you do? You drink, you drink, you drink, and you drink. You drink maybe 10, 15, 20, 25 drinks, right?"

Ah right, yeah.

"So what happens is you get all this booze in your stomach and you start getting high…

Do you hear…

… "as the alcohol"…

Do you hear those rounds?

"Yeah, I do. Mortar rounds."

They're flares.

"Oh yeah, we're not worried about that. We're going to go take a cold shower. What happens is you get all this alcohol in your stomach and it's absorbed by the bloodstream slowly. So if you were to come back here and lay in the rack, what would happen is for the next two or three hours, you would just get drunker and drunker and drunker and feel worse and worse until about midnight when you begin sobering up. While you're sleeping, you're not using up any of that alcohol in your bloodstream. So you wake up the next morning with an Excedrin headache."

That's right, an Excedrin headache.

"So the way you get around that is you drink fast during the game. You drink and drink and drink and get your stomach all filled up with booze. And then you come back to the hootch area and you swallow your finger. You barf up all that extra booze that's in your stomach that hasn't gotten into your bloodstream yet. You barf all that up and that way you won't get any drunker than you already are."

That's Tom's method.

"Right, so as soon as I get back in the rack, I'm as drunk as I'm going to get tonight right now. As soon as I get in the rack I start sobering up. I take two Excedrin to get rid of the headache in the morning and I'm all set to fly. That's the purpose of barfing right now so you get all that out of your system."

All right, now what you all just missed was young Tom and I went out and swallowed our fingers. Young Tom barfed, but the Ace didn't. For some reason, I just could not barf.

"You're a pu**y. You didn't swallow your finger far enough."

I'm not a pu**y.

"You have to stick your finger way down your throat and tickle way down in your fallopian tube. And when you get that tickled, then that's when the guy inside says, Jesus Christ, they're calling me, I'd better barf. So you barf right there on the street."

Hey, I don't have a fallopian tube.

"You got it just like everybody else, a fallopian tube right down in your throat."

That's a negative. Nah, I don't know if you realize it or not, you dumb shit, but only females have fallopian tubes.

"That's a negative. That's a ueskit tube...eustachian tube. That's right, you asshole. Only females have fallopian tubes."

That's your esophagus.

"...right there in your, your usopatube is right down there."

Folks, you don't realize what a dumb shit Tom is until you hear him talk.

"But I can sure play Whales' Tales."

Yeah, he's good at that. Now we also played buzz tonight, but he didn't do too well on that, dumb shit that he is.

"That's bullshit. That's not true. I did great."

Well, who didn't do great, then?

"Barry. Barry and Theodus."

Why are you drunk if you didn't do good?

"I'm not drunk now because I swallowed my finger and I got all that extra booze out of my stomach. You had a lot of sympathy and a lot of team drinks."

Now you're telling me, a lot of sympathy and a lot of team drinks. That's Tom. We sat out on the patio tonight and the war was going on while we sat there and drank.

"Why?"

They're launching all the Hueys out and they're going to...well, they've been shooting mortars and they have flares going off over there. Now you hear that, folks? They're launching all the Hueys. They got flares going. They got some g**ks cornered because, after all, this is Tet.

"You know where those g**ners are?"

No, Tom. Tell me.

"They're right down from the perimeter from where we were sitting at the club, right there in the goddamned wire. We were sitting there playing college games on the patio within 200 meters of the perimeter where the g**ners are now and they're shooting at us."

Aren't we brave?

"We sure are. We wouldn't have been there if we'd known about it. It…"

Unless we had been drinking.

"But we were drinking, weren't we?"

We sure were.

"But now you see I've swallowed my finger and I'm all set to ring the bell for freedom."

Right.

"But you're not, because you didn't barf. You're going to get drunker and drunker and drunker as the night goes on."

"Oh, I do."

You do, that's true. And so that's why you have to be sober.

"I don't have to be sober. I can do like Stubby did and pu**y out of my hop."

Yeah, Stuber pu**ied out. He barfed all over the ready room.

"It was disgraceful."

It was. It was very disgraceful.

"And then he came back here and he stole my pillow and he went to sleep."

That's right. That's what he did.

"Yeah, but here we are on the 17th of February. What the g**ks call Tet, tit, teat, what…"

That's right. It's Tet. T-E-T.

"Tet. Tonight is Tet. Boy I wish I could get my hands on a tit."

Let's try to keep our minds on the war. There are bad people running around here trying to shoot us.

"And how would you feel if they shot me?"

Shot you? I'd feel sad, real sad.

"You wouldn't do anything about it, though. You'd go hide."

I would hide, that's right. What would you do if they shot me?

"I'd hide."

You're damn right, you'd hide. And do you know why? Because you're a pu**y, that's why.

"How can you say that after all the times I've defended you?"

What do you mean, defended me?

"Like all the times I stepped forward to defend your honor."

Like when?

"Like the time Bob called you a pu**y. I defended you."

I guess you're right.

"Let's go take a shower to sober up. To hell with the war. I have to fly tomorrow, the cold water will help. To hell with the war. I'm too short to worry about the g**ks."

Okay, folks. Tom and I are going to say to hell with the war and go take a shower…together.

"This is…if it's Tet, it's a holiday for them. So it's a holiday for us, too. We don't get many holidays, right?"

Right. And when Tom and I get a holiday, we take a shower together.

"That is one thing we do, we take a shower together. Ace is a good shit. You've got a good son, Mrs. Mother Ace. He's a worthless piece of shit, but he's a good guy. But don't worry. We only shower together on holidays."

Okay, so listen. We're going to sign off now because we're going to go take a shower with a friend. But I want you to know we'll be back.

"We'll be back. We surely will."

And we'll talk again.

"After a station break."

February 18, 1968

Hi folks. Well, here we are at Phu Bai. I've been getting a little sun, the first chance I've had for that in a long time. I need to get a tan before I come home, I guess. Also, I need to get my body glued back together after last night. We sort of had a party last night.

Tonight I've got the mortar watch, the first shift of it. All I really do is launch and go up and fly around for about two hours looking for mortars, you know, and any hostile fire. As far as Tet is concerned, it's been pretty quiet. We haven't had much problem here at Phu Bai, and are pretty happy about that. You'll recall we said last night we had heard they had some g**ners trapped, but I haven't heard any more about it. I don't know if it was true or not, and don't really care all that much.

On the 25th, the skipper is going to give me a day off so I can go down to Da Nang and see Sabine, the German girl I've told you about, before she goes home. I think she's supposed to leave for Germany on the 26th, so we made plans to get together before she leaves. Things like this make the war a little bit easier to take.

Well, I can't think of anything more. We really had a good time last night. Well, I'll talk to you again.

February 21, 1968

Well, not too much has happened in the last couple of days. I flew yesterday, carried passengers for four and a half hours between here and Da Nang. Some of the other guys went up and worked in the mountains in the north. I guess you probably read in the papers that the U.S. has gone back into the A Sầu Valley. Well, of course, where the grunts go, the aviators are soon to follow. So we're back in there supporting those guys. It's not all that good in there, actually. It's not like in the south where we're used to working. I prefer the south to the mountains.

The day before yesterday I led a section into an area south of Marble to do some work. It just happened that one of the loads that we were supposed to drop off went to a place about 500 yards from a hospital, which happened to be that German hospital near An Hoa. So of course I stopped in to visit and got things worked out. Anyway, I'm looking forward to getting down there to see her.

Phu Bai is not all that good. It's dusty as hell, just unbelievably dusty. Except today it rained so it's muddy instead of dusty.

I got a letter from my brother, believe it or not, and I'm going to try to whip off a reply to him. He said he's thinking about coming into the Marine Corps, and I've got to tell him he doesn't want to do that.

We've moved from the first place we lived in at Phu Bai. The first place was kind of a rat trap. It was one of these Southeast Asia hootches that consist of a floor and a roof with screens for walls. Boy, was it hot. Now we've moved into a big Quonset hut that is pretty nice. It's kind of partitioned off, though the partitions consist of blankets and wall lockers. We have a back room with a refrigerator and a couple of hot plates. It's sort of like our reading room; everyone congregates back there at night. It's really pretty nice, probably the nicest hootch I've lived in since I got to Vietnam. And we have a hootchmaid.

Which reminds me, we had a hootchmaid at Marble Mountain and when we left we gave her a Christmas present, which seemed like the thing to do. But we have discovered we have been assessed a Tet bonus for the hootchmaids at Marble whom we haven't seen since the latter part of December. It's really pissing a lot of people off. It sort of reminds me of the government at home, the way it takes money away from the people who work and gives it to those who don't. So a lot of us have just flatly refused to pay. I don't know what will come of it, but we already gave our hootchmaid one gift, and I don't think they should make us give another one.

Well, all for now.

February 24, 1968

I have a little sequel to the story about the money for the hootchmaids. Our skipper called us together the other day to inform us we would pay the $4. It was a mix-up on the part of the group to just go ahead and pay these people this Tet bonus and then assume they could assess everyone to make up the money. But since the skipper has been directed by group to pay a certain amount of money, we had no choice but to back him up. Otherwise he would have to pay it all out of his own pocket. Well, enough of that.

I flew the mortar watch again the night before last. It wasn't all that good. The ceiling was down about 500 feet, but they had a high threat and it was the night Da Nang got hit so badly. They were afraid we might get hit too, so we had to launch despite the weather. It was kind of funny. I got out there and was flying around in a bounce pattern because I was afraid to get outside the field boundaries, afraid I'd get shot up. So I just sort of stayed right inside the boundaries doing touch and goes. Then the bunker called me up, the command bunker here at Phu Bai, and asked what the ceiling looked like. Well, I told them there was a broken layer about 500 feet, a pretty heavy layer broken in spots. So the guy says, "Well, how about on top of that?" I said I didn't know, so he suggested I go up and look. And I said, well, how about if I don't. I was afraid if I got up above that layer I might not be able to get back down through it again. So the guy said okay and I spent the rest of the night flying around at 500 feet from 3 a.m. to 5 a.m.

Today, interestingly enough, I took over as the S-3, or operations officer, for the Squadron. It's kind of a joke really, but they've been breaking me in the last couple of days. It's only until the Major comes back, not a permanent thing. He left for a week or two. But I will be the Operations Officer for our next move back to the ship, which should be kind of interesting.

Tomorrow, tomorrow, tomorrow. Tomorrow I'm going to Da Nang to see my honey, Sabine, the German girl. I'll go down sometime in the morning. I hope to stay at Marble and come back sometime on the 26th. I am hoping everything works out okay without any hitches. Mainly I'm worried about getting out of here. The weather has been real bad the last few days, real bad. I just hope it improves enough for me to launch tomorrow.

I haven't been doing a whole lot other than playing poker. I started playing the other night, got in a game and won $17. Then the next day I lost it all back. It's just something to do.

I don't know if I told you, but I got the packages from you and Betty. I appreciate them. You can pass the word that they were received and went over real well. Also, tell Mrs. C. that her fudge was a big hit with the guys. Everybody really liked that.

Well, all for now.

Chapter 17
Questions from Home

Hi folks. This is still the third of March, but I just got your tape and thought I'd try something a little different; I'm going to try to answer your questions as I go through the tape. I've gotten part way through now. I got to the point you mentioned Major M., and I'm going to try to respond to the others by memory.

First, yes, I knew Carolyn A. She married a guy named Perry, I believe. I can't remember his first name, but she used to live out there in the first house on the left before you get to the corner by the McNeill's place. I'm sorry to hear that her kid got killed, but you can probably understand that I'm not too upset. Death is so commonplace over here, so many people dying.

About Major Ribbit, I'd be interested to know where he is. I might drop him a line to see how he's getting along, and see if he remembers me.

About not mentioning Anna Faye and Larry. I guess I haven't been too good about it, but I have received everything they sent. In fact, I got a picture that Larry sent me that I keep with all my stuff. You know, my more valuable things that I keep in one spot. I've got that and the darts Larry sent me. They go with me everywhere, aboard ship and here to Phu Bai. We've got a lot of pleasure from them. We do a lot of dart throwing, sometimes for money. But I appreciated all the things sent to me, and I'm sorry if I haven't acknowledged everything, I really am.

Now about Major M. He is not in Da Nang, he's in Phu Bai. I've seen him several times in the club. In fact, on many occasions he and I have had a few drinks together. I mentioned to him the thing about Dave's graduation, but there really isn't a lot he can do. He advised me to talk to my skipper, and I did. But there's not a lot he can do either, what with my time coming up and us being short-handed and all. I mean, I could make a federal case of it, but we're short on people and they need me. So if anything happens, it will have to come from the other end.

Mom, you and Pop should get together and confer on the jokes. I think the story about the rabbi and the priest somehow got on the tape twice by different people.

And now, about The West Virginia Hillbilly. We were curious who got that subscription for me. Yes, they come once a week. In fact, I got a copy of it today. I sort of thumb through them and leave them in the ready room for the other guys to see. They have produced quite a few laughs because when the first one appeared, there was no debate at all about whose it was. So yes, I've been getting them and for a while I wondered who was responsible, and now I know.

Yeah, I remember Donny K. He used to sit around in the Student Union there at Marshall and talk to us. He was a pretty clean-cut young lad, as I remember. I was interested to hear about the change in his appearance since I left Marshall. But he was always a decent guy, and I would imagine he still is, even though his appearance might suggest otherwise.

In the event you hear from Dan Mc., the guy who went through Pensacola, I was interested in what you said he said about coming to Vietnam, the part about him hoping when the time comes he can carry off his duties with as much courage and lack of emotion as we do. Please tell him it isn't courage.

We do what we do because we have to do it, not because we want to. As for emotion, I can guarantee you there isn't one guy up there who's not scared, and it's about all we can do to talk on the radios. So there is no lack of emotion. We just try not to let it show, but it's there.

And please thank Aunt Sue for the books. It's nice to have something like that to pass around. A lot of the guys besides me read them, and we all appreciate them.

You asked what I would like for my birthday. I will tell you. I would like to come home.

You mentioned my roommate in Florida, but didn't say which one. I'll just have to whip through them for you. First, there's York who is right now working as FAC, or Forward Air Controller, up in this Operation Dewey Canyon with the 9th Marines. He has been seeing some serious shit of late, lots of enemy action. Dave T. is here at Phu Bai flying 53s and I run into him occasionally. He's doing okay for himself. The fast man, Bill S., is down in Da Nang now. He was flying backseat of Phantoms, you know, the F-4s, and has since gotten himself a desk job. I'm envious of that. So he's doing pretty well for himself. As for the round man, Bob L., he crashed his Huey here. It was a long time ago. He wasn't hurt, got out without a scratch. In fact, everybody got out unhurt. So they went back to Quảng Trị that night and the round man got drunk, fell out of his bunk, and broke his leg. They had to MedEvac him out of the country. Right now he's in Okinawa in charge of the NCO leadership school, which is not a bad deal at all.

And the other guy, Chuck M., he's a Communications Officer up in the northern I Corp area someplace. I'm not sure where. I ran into him one day down in Da Nang, but I didn't get to talk to him very long. Anyway, that about covers all my old roommates, all I can think of at any rate.

I noticed in your tape here that you made several allusions to Jerry S. and the comments he has made on past tapes, mainly the "Hi Johns" he throws in here and there. So I played that part for Jerry and now let him speak for himself.

"And a great big hi, Mrs. Ace. I feel kind of guilty now because of all the comments you made on the tape about hi John. Actually, the only reason I tease him is because I think it sounds kind of neat. And John is just sort of easy, you know and we all get a kick out of teasing him.

But seriously, we all really do enjoy your tapes. I enjoy them because I'm not from that part of the country where the people have an accent, and I get a kick out of hearing all of John's friends, especially some of the gals. It really shows up in their voices. Just a different way of talking that people from the hills have, I guess."

Thanks, Jerry. Now Mom, we assume the 'hi Jerrys' we get are in response to the 'hi Johns' you get on the tapes I send you. Aunt Sue mentioned that you all were fixing up the house and that I would enjoy it when I get home. Tell her I would enjoy it if we were all living in mud huts.

Which brings me to Christmas. Now don't go planning anything real big for me when I get home in July. Just getting home will be treat enough for me, believe me. Don't go to a lot of trouble and get the house all decorated and all that stuff, because I'll probably be flitting from place to place. I'll find it hard to stay in one spot, I'm sure. So don't go overboard with it. It wouldn't really make a lot of sense to have Christmas in July. And you can get me a present if you want to. But as for me, getting home in one piece will be present enough.

As for the German girl, she will be going home in a day or two. The skipper gave me a day off on the 25th of February, which was supposed to be the day she would come into Da Nang from An Hoa. So I went down there and ended up spending two days there because they forgot about me and left me there for another day. And it was the third day before I got back here.

But she never made it. The area around An Hoa had been hard hit by rockets and mortars, and also a coordinated ground attack. So I assume she was pressed into service working on the wounded, and I couldn't wait around. I had to come back up here to Phu Bai. But I have her address in Germany, and I will write to her.

You mentioned that article about the SDS, the Students for a Democratic Society, in Saga Magazine. Well, we saw that article. It's a pretty radical group as near as I can tell, dedicated to anarchy. So you can assume that most of us here agree about the SDS. Also, I saw that article in the magazine about the

Kennedy assassination. It really does make you wonder if there wasn't something else going on, more to it than meets the eye. It will be interesting to see if that prosecutor in New Orleans, the Garrison guy, comes up with anything.

You can tell Mrs. C. and Mrs. J. that I did not get to attend the Bob Hope Christmas Show. I spent that day flying. I was carrying mail and food to troops in the field during the morning, and in the afternoon I took a chaplain out in the field to conduct religious services. So I guess if I want to see Bob Hope, I'll have to come home and watch him on television.

Tomorrow I'm supposed to go back north again, and I am not real thrilled about it. We were supposed to go back aboard the ship but things changed again and now it looks like that has been put back. So I should see the DMZ again tomorrow. As I said, I don't like to fly up there. That place just scares the hell out of me, mainly because of the low ceilings. You don't get the terrain clearance you really need, and also that is definitely g**ner country up there. You don't want to mess around because those people will just shoot the shit out of you. Or as one of our friends recently put it, they will put the super shoot on you.

Those guys up there in the north don't screw around with small arms fire, like they do down south of Da Nang. Down there when you get shot at like nine times out of 10 it will be 30 cal, you know, rifles and things of that nature. But up in the North Country, if they even take the trouble to shoot at you, it's with 50 cals. And down in the south, if we get shot at with a 50 cal, it will only be one 50 cal. But up north they have dual mounted 50s and quad mounted 50s. And they won't have just one in a position, they'll have a bunch of them. And they are good with 'em, too. When they cut down on you, they'll flat put a hurtin' on you. It's just kind of scary. To be perfectly honest, I don't really care for it all that much.

All the same, the trip up there is kind of funny. You've heard of Highway 1 which runs north and south here in Vietnam. It used to run all the way into North Vietnam, I think, but now it just stops at the DMZ. But anyway, to get from Phu Bai to Quảng Tri, you have to drive or fly up Highway 1, and you have to fly at low level because of the weather conditions. It's interesting to see it. A lot of places along the route of the old railroad track is still in place, and the Vietnamese have built their homes right on top of the tracks.

It's an interesting flight. We go past places like Đông Hà and pass on into Quảng Tri, turn west and go past a place called Mỹ Lộc. Then we go past a place called Cam Lộ, and up there we look down towards the south and see the remains of Camp Carroll. Then you drive on past the Rockpile, hang a left, and drive down the valley to LZ Vandegrift Combat Base which is now a giant re-supply or staging area for all of the re-supplies that go out into that area, the various battle zones of I Corps.

The operations there in the mouth of the A Sầu Valley have actually extended over into Laos. I don't know if this has been publicized, but we've had birds go on MedEvacs as far as six kilometers inside Laos. So I know we have people in there, and apparently they are there in some strength. But the g**ners have all kinds of brass up here and have built a road that comes down from North Vietnam, cuts through Laos, and comes into the A Sầu Valley. They're just brazen as hell about it. They build it at night and hide during the daytime. They drive trucks along the parts that have been completed. Our grunts actually hear their trucks at night. We've captured some of their tracked self-propelled guns and large artillery pieces. They have a lot of equipment in their caches. It's all new gear and apparently hasn't been there very long. There's a lot of it, too. So apparently we've hurt them pretty bad, although we didn't take all that stuff cheaply. We took a lot of casualties.

I don't know if you've heard about Firebase Russell. It was overrun not long ago and they took something like 38 KIAs. The place was overrun at night. The g**ners used rockets and mortars to support a coordinated ground attack, and the fire base was unable to hold them off. I flew over that day before yesterday, and it was interesting to see. It had just been blown to shit; I couldn't believe it.

That's another thing about the whole area up there just south of the DMZ and the DMZ itself. It's just had so many bombs and artillery rounds landing in the area that the entire region is just devastated. It's truly amazing. There must have been simply millions and millions of explosions up there. It really shows up in the face of the land.

My mission tomorrow is stand-by MedEvac. I'll be making two scheduled runs from the hospital at Quảng Tri out to the hospital ship. I don't know exactly what I'll be carrying, but probably patients who have been stabilized to the point that they can be moved to better medical facilities. Also, they perform autopsies out there. We took Tom B.'s body out there yesterday for an autopsy, along with another boy. We went out there and had to send in two people to identify the body. The Squadron flight surgeon was with us. I had no great burning desire to go in and see Tom, so I sent my crew chief. I mean, it's not so bad when it's people you don't know. But when it's someone close to you, it preys on your mind. You try not to let it, but it does.

Well, it's after midnight and I have a 7 a.m. brief, so I better get to bed.

March 5, 1968

I went up north again today and did some re-supply to an Army LZ. I enjoy working with the Army; those people have their shit together. They were all up on the radio, particularly the people in the initial pick-up zone, the staging zone. I only had to call them one time. When I took the first load out into the field, everybody on the tactical radio net was up and waiting for us to come into their zones. They had the smoke ready to pop, they had their zone briefs prepared and it made everything work very smoothly.

Prior to that, I did some work for the 11th Engineers. I took a party of them to LZ Vandegrift, or Vandegrift Combat Base. Maybe you've heard of it. Anyway, I went up there, shut down and waited for them for about an hour. They're lengthening the runway there to make it long enough to accept Fixed Wing and C-130s. Right now, it appears they're starting to withdraw Marines from the A Sầu Valley. I think they're going to start putting them back northwest of Khe Sanh Rockpile and areas in there.

Anyway, the g**ner activity just south of the DMZ has increased in the last couple of weeks. In fact, today right close to a zone that I was supposed to go into to re-supply, we had a bird going in for a MedEvac take 40 hits. One of the pilots was wounded. It's pretty exciting. I was glad later that they had cancelled the re-supply to that particular zone so we didn't have to go into it.

I've been afraid I'm going to have to lead a section up there on MedEvac, and I'm not really excited about that. I led a section today, but it wasn't too bad, the areas we were working in were all relatively flat. It wasn't too difficult to find our way around. But I have no desire to go back up there and drive around in the mountains because I don't know the area and have no idea where I am. It's difficult to tell from the maps, sometimes. Also, the mountains are so high it's difficult to get TACAN bearing and distances from places like Quảng Tri and Đông Hà. And with the low ceilings, it's hard to maintain proper terrain clearance to stay clear of ground fire. In a lot of cases, there are a lot of g**ners very close to the zone so it's difficult to get in and out without getting shot up. So I'm not really excited about the prospect, but the time is coming when I'm going to have to do it, although I will continue to try to avoid it.

The ship came back the day before yesterday, but the guys on that ship apparently aren't going to come up here. We're kind of pissed about that because the guys on the ship have had about two and a half weeks of free time, now. They were in the Philippines for two weeks with nothing to do, and we've been carrying the load for the Squadron all along, which hasn't been a lot of fun. We would like to see them come in and get some of the action, to spread the wealth around. This is not the kind of thing we want to be selfish about.

Well, I don't know. I haven't been shot at since December. January was a good month. February was pretty good, too. And I don't really relish getting started again because I never did enjoy being shot at. It's not a lot of fun. As a matter of fact, if you think back to something you dreaded doing, something you really did not want to do, and you multiply that by 100, you can get an approximation of how I feel about going back out and getting shot at again. Well, I'm hoping things will work out.

We have the old Alpha models so we can't do very much back up in the mountains. The birds just can't haul heavy loads into those high zones, so they've been using us for MedEvacs and VR missions. Well, I can't think of anything else right now. Later.

March 7, 1968

Good news, folks. Tomorrow we go back aboard ship. Everybody is quite happy about that, although we don't know exactly what we will be doing, whether we'll be working off the ship in the north and doing about what we're doing now, or whether the ship will go back down to the Da Nang area. Most of us aren't really excited about flying back around Da Nang because it's been so hot down there the last few weeks. In fact, it's been hotter lately than it has been in the whole time we've been over here, which is pretty hot. We have no great desire to part of anything of that nature.

I got a letter today from Sabine, the German girl. As you recall, she never showed up to meet me in Da Nang, and when I got back up here, everyone said I had been stood up and shit upon. Well, I never really believed that and now I got a letter that confirms what I thought. They had changed her departure time and everything, leaving her unable to meet me as planned. So now I feel a whole lot better about it. As a matter of fact, I decided I'm going to Germany next summer after I've been home a week or two. The guys have been kidding me, saying this is the most expensive letter I ever got. Well, of course, going to Germany will cost some money. But I think I should have enough saved by then to cover it without any undue strain.

By the way, I have a cold or something. I've been sneezing a lot and my head is all stuffed up. Well, a buddy of mine was telling me he's got orders for Quantico, something I would like to get very much. My orders should be coming through in the next two weeks or so. As soon as they do, I'll let you know what I'll be doing when I get back to the good old U.S. of A.

Well, I have to get busy packing now and get everything ready to go. So I'll try to talk to you tomorrow when I'm back aboard the ship.

March 11, 1968

Well, here I am aboard the U.S.S. Valley Forge. We've been here a couple of days now. This is the first chance I've had to sit down and talk to you. Things have been pretty hectic. I've been doing a lot of flying. I went out today, but I should have just stayed in bed. I was on the schedule for an admin run. There was supposed to be an 8:30 brief and a 9:30 take-off. Well, about 8 a.m. they gave us a call and told us we had to launch at 9. So I get dressed, go down to the ready room to get the brief. Only problem was they didn't have it yet. Nobody knew exactly what we were supposed to do. So we fiddle-farted around trying to find out what we would be doing. I was flying in the wing bird today. The guy from Rising Sun, Maryland, named Norm Kirk, was to lead the section.

Anyway, we finally got things pretty well together. So we walk up to the flight deck. Now, here on the Valley, the flight deck is five decks above the ready room which means we have to climb five flights of stairs, whereas on the Tripoli the ready room was the next deck down from the flight deck. Well, anyway, we shouldered our gear, charged up five flights of stairs to the flight deck, got up, signed off our planes, got in, and started turning them. All of a sudden Norm's airplane goes down. Well, Roger, okay. I'm going to shut down too and wait to find out if they can fix his bird or have to get another airplane. Then all of a sudden he starts turning up again, and so I did too. But then his went down again. Well, in the meantime I developed problems in my hydraulic system. I had no pressure at all and I couldn't get it up. So I had to down my bird. Now we've got two downed airplanes, and we're running short of airplanes.

So we take off all of our gear, trudge back down to the ready room. In the meantime, they get Norm's bird back up. Well, there was another guy on the deck doing a turn-up, so they decide to give us his bird so we can continue with our mission. So we trudge all the way back up to the flight deck again. But just as we got up there, the other guy lifted off the flight deck and left. So we took off all of our gear and trudged back down to the ready room again.

Now everyone is totally confused. No one knows what is happening. So finally they decide I will be standing by.

Okay, great. We're going to be standing by. So we stand by and we stand by and we stand by. So finally about 1 0'clock they say we're going to launch. We're sitting off the coast of Da Nang, right? So we're to go up off the coast up near the DMZ to the hospital ship, pick up a load of blood, and bring it back to the other hospital ship in the harbor. Okay, no sweat. So we get up, climb up to the flight deck, go out to our airplanes, turn them up, and Norm's goes down. So I shut mine down. We go back down to the ready room. Then we find out they want me to go single plane. Okay. So I pick up my gear, trudge back up to the flight deck, turn it up, and leave.

Well, I got just past Da Nang harbor, out near a big mountain that comes all the way down to the sea, when all of a sudden the weather just turns to shit. I mean literally. The ceilings are down to 200 feet, the visibility is about one quarter mile, it's raining, the wind is blowing, and fog is all over everything. So I drive in this for about an hour. Did I mention the turbulence? Anyway, about half way through, I lose my TACAN. So I had to just go along the coast and sort of follow the beach. I finally go up to the Cửa Việt River where I found the hospital ship. I call them up and tell them I'm inbound to pick up the blood. So the ship's tower says Roger, okay, wait one. So I wait and then they come back on and tell me the blood has already gone by boat.

Well, as you can imagine, I was royally pissed off to have come all that way through that for nothing. I started to turn around to go back, but then they said that since I was already there anyway, I could pick up some passengers and some mail. So I picked up the load, stopped off for some fuel, and drove all the way back to Da Nang through that soup. I took some people to the hospital ship and then took the mail over to the naval base. I got to the NSA hospital and dropped off some more passengers, come back to this hospital ship and pick back up one of the passengers I had dropped off there, and head back for the Valley Forge. So I call in and tell them my mission is complete, so they say why don't I drop in the landing zones around Da Nang and see if there is anyone out there to pick up, so I head on out, pissed and tired.

So anyway I get into Marble because I need some fuel and I see three Yankee Tango birds sitting in the fuel pits there. I tried to find out what they were doing in Da Nang. I thought maybe they were picking people up but they weren't. So I lift out and I go over to Three MAF, pull up to land, and I see another Yankee Tango bird sitting there. So I say to myself, well, I'm going to wait and follow this bird when he goes out to see if he goes into any of those other zones out there. Then I'll know he's an admin bird and I'll just forget and go on back to the ship. So he lifts out and I follow him and, sure enough, he starts an approach to another one of the zones out there. So I figure, yeah, that's the admin bird. So I pull in power to leave and head back to the ship when all of a sudden I lose one of my engines. Well, okay, now I've got real problems, and my co-pilot was sort of asleep on me. I yelled at him three times that we'd lost an engine and I wanted him to pull a condition lever off. The temperatures were just shooting sky-high and I wanted to get the engine completely shut down to keep from having a fire. I yelled at him three times and he still didn't hear me. I finally had to hit him on the arm to get his attention and get him to pull the condition lever off. I couldn't do it because I had my hands full with the collective and the cyclic and getting everything all settled up because we had a load onboard. Also, we had a heavy fuel load. When you lose an engine like this, your turns start to decay, your rotor RPMs. They eventually decay to the point that you can't fly anymore. So I had to get all that stabilized out. We finally got it into Marble. I had the crew chief check it over, try a relight on it, and it started right back up. Just amazing.

But in the meantime, this other admin bird had seen me leave the place where he was entering, and saw me dumping fuel. So he decided to follow us to see what the problem was. So I put my passengers on his plane and I came back to the Valley Forge. And would you believe they had 53 knots of wind across the deck. That's a lot of wind. You figure about 1.6 miles per hour for every knot. I guess it's a little over 60 MPH I'm trying to land in. That was truly hairy, and that was my day today. I'm hoping tomorrow will be a little bit better.

Although I'm sniveling tomorrow, I've got to take the instrument exam. Every year, we have to requalify for our instrument cards, and I have to take the written exam. After I get that out of the way, I have to take the flight exam. I have to get all that done sometime before my birthday, which leaves me about 20 days to get it done.

As far as the war is going, I'm not real sure. We got out of the north, and I don't know now if I'm happy about that because now the south is getting hot. In fact, it's hotter now than since we've been here. It used to be that we went to An Hoa if we had 1,500 feet. Now if you can't get 2,000 feet of terrain clearance and stay below the clouds, you don't go. The people out there are becoming more efficient and they are getting better weapons.

You know, I've noticed something interesting over here, and it's something a lot of other people have noticed. The Army goes into a place and they thrash around, and they end up getting it secured so you can drive around without getting your ass shot off. But the Marine Corps goes into an area and they thrash around and thrash around and thrash around for years, and the situation remains basically the same. Sure, they kill a lot of g**ks. But they never seem to do a whole lot toward securing the area.

I think part of the problem is that we're not using the Marine Corps for the purpose it was designed for. The Marines were not designed as a holding force. Our supply system is not set up to handle a protracted land war. We're organized to make initial strikes and, you know, get right in there kicking ass and taking names. But when that phase is over, the Marines should be pulled out so the Army can come in to organize and hold the area. Now, when the Marine Corps has been allowed to fight it's kind of war, they've done an outstanding job. But when they're forced to hold, as they have been in the I Corps area here, they just don't really get the job done.

Well, it's getting kind of late and I have to get up tomorrow. So I think I'll close off for now.

March 14, 1968

Well, another week is out of the way and I'm serving as AM test co-pilot which means another day of no-fly, no-die. I'd like to clarify a few things for you. First of all, when you hear about boats being sunk in Da Nang River, and Da Nang being rocketed and mortared, please don't worry. The Navy thinks a lot of its ships and at night we pull well out to sea, probably 15 or 20 miles out. There's not a whole lot of danger. The only danger comes with flying in-country.

I went out flying wing on an admin run yesterday. My leader managed to take some fire going into a zone. It looks like we may be getting back to the shooting war here pretty soon, and I'm not really psyched up for it, to be honest with you. I kind of like flying admin runs and sort of messing around.

But in the next couple of weeks, we're supposed to backload the battalion and the battalion is in contact right now. They've been in contact ever since they went ashore. So it looks like when we go to backload them, we'll probably see some of the shooting war again.

And then, after a couple of days of refitting and everything, we'll be taking them back out to the field, and I'm not looking forward to that. Once we get them out, we'll be required to support them directly. We'll be doing their MedEvacs and re-supply. So I'm sure we'll see some more of the war.

Now Mom. About you not being notified the two times I was injured. On the form record of emergency data, I requested that you not be notified in the event of injury. The only time you'd be notified is in the event of my demise because I figured in the case of an injury I would rather tell you myself. So excuse me. That's why you didn't hear about that.

Now Mrs. Gilbert asked about her friend's son who is going to be a Navy Corpsman, and I think Ron told you what he thought about it and I agree with him. The guys have a hell of a job to do particularly here in Vietnam because they're required to put out a lot more on occasions than the average Marine. A lot of times in a combat situation they've got to get out there and expose themselves to risks far greater than the average rifleman in that they have to treat the wounded on the spot. I tell you my hat is off to those guys. I've worked with some of them on MedEvac missions and seen some of the work they do. These guys will jump right out the back into a hot zone and start working on MedEvacs, just ignoring the lead flying all around them. They don't seem to care much about their personal safety. And I don't want to get anyone upset, but you can tell Mrs. Gilbert I certainly would not want to be in her son's shoes. I mean, what we're doing is bad enough, but what they're doing is infinitely worse. Of course, not all of them are working in the field. But the ones who are, boy, are really hanging their nuts way out.

Well, I want to talk to you some about what we'll do when I get home. But I see this tape is just about done. So I'll wrap it up for now, and talk to you more about that stuff on the next one.

Chapter 18
Getting Shaky

Hi folks. Today is the 17th of March, Saint Patrick's Day. Well, things are going along pretty well, I guess. Not much has happened in the last couple of days since I finished that last tape and sent it to you.

I got a letter from Virgil the other day. He said John F. would like to use our cabin as a guest cottage. He offered to trade us a lot of our choice for the cabin. I told Virgil to forget it, that it was a bad deal all around. Those lots are going for about $3,000 now, while we're paying over $6,000 for our lot and cottage. So that's a rip off. I don't know why Virgil even considered the deal; maybe now that he's married he feels kind of saddled with the place. So I don't know. I told him if the offer was two or three lots, maybe I'd consider it.

We're getting ready to go back up north near Quảng Tri, off the coast from the DMZ. I'm not really excited about flying in there, even though that area hasn't been as hot lately as it has down around Marble Mountain.

Apparently, this ship will be there until it's time to go to the Philippines. We're going back for replenishment, though I'm not real sure what that means. They stay on the line for about 45 days and then go back for about 10 days, and then return to action. So I should get to go with them when they go to the Philippines this time. It should be a good deal for me, and I'm looking forward to it. Anyway, if rumor holds, we'll be up there in the north until the 19th or 20th, and then head for the Philippines.

Today we took hot chow out to the field to two companies that were out there. We got it in okay, and when we went in we had gunbirds working with us, and jets running airstrikes. They were putting in napalm, and that is truly a thing to see. It's just beautiful. You know, it makes me think about some of those papers you sent to me, the ones put out by the SDS knocking napalm. If the people who wrote that were to get in a situation where they're getting their asses shot off, they would appreciate how truly beautiful napalm is. It does a much better job of killing than bombs do. A bomb can hit close to someone but if they're in a bunker, or some other defensive structure, it won't hurt them. But napalm goes right in there and does a job on them. If they're in bunkers, it will smother them. If not that it will burn them. It hits the ground, flows around and sticks to things like people and burns them badly. It's kind of nasty stuff, really. The results aren't very pretty. But it sure is great in terms of discouraging people from shooting at you.

Also, it's beautiful in terms of colors, the explosion and everything. I hope I can get some good photos of it so you can see. As far as I'm concerned, Dow Chemical can continue making napalm just as long as they want, because I, for one, am all for it.

Anyway, we went in and had the gunbirds to take care of us so we weren't really too worried, didn't have to sweat it much. They had taken some small arms fire in one of the zones, but I didn't go down in that particular zone, so I didn't have to sweat it. We went in, dropped off the hot chow and passengers and lifted out. We came back on a passenger run to the ship, and went back out about an hour later to pick up the empty cans and the people who had served the hot chow to bring them back to the ship.

On that last trip the gunbirds were gone so we had to do everything ourselves. So I called the people on the ground and told them I was inbound into their zone and wanted a zone brief. So the guy comes on the line and says, Roger, he had friendlies out to the north about 200 meters, and friendlies out to the east about 200 meters, and friendlies out to the west about 300 meters and friendlies to the south about 150 meters. Also, he said they had a listening post 400 meters to the east, and another one someplace else, I can't recall what all. I don't know. Anyway, he said they had taken negative enemy fire other than snipers in the last 24 hours, and suggested I make my approach from the southwest. So I said Roger, I copy your zone brief, and if you'll pop a smoke I'll start on down. So they popped smoke and I started down. Then the guy added, sort of like an afterthought, oh by the way, don't fly over the northeast of the zone very far because there are some g**ks out there somewhere, and we think they may have a 50 caliber machine gun. Well, at that point, I almost decided to pack it in and pull back up to 6,000 feet. I couldn't though, because I had already started down. But I got into the zone, picked up the chow cans and got back to the ship okay.

Oh yeah, listen up. I meant to tell you that the other day for lunch we had lobster tails. Now how bad is that? Shipboard living is okay, it really is. The only problem, and one that doesn't really bother me a lot, is that they have no bar on the ship. Now a lot of guys have bottles they brought aboard with them, and they drink on the ship, but I don't do that. It seems to aggravate my seasickness. Just the smell of it can turn me green, so I'm one of the abstainers on the ship.

Otherwise, the ship isn't bad, not bad at all, and the store is pretty well stocked. I've been looking at watches, particularly the Accutron watches with the tuning fork inside, the Bulova watches. But I haven't decided if I want one bad enough to pay all that money. I'm still kind of saving my money for Germany next summer. I've got almost $1,000 saved you know, so by the time I get home I should have the right amount; enough to go to Germany and still have some left over.

Oh yeah, by the way, I had a tape that Gene C. and I were sending back and forth, but when Gene left his job there at the radio station he no longer had access to a cartridge tape. So I used that same tape to record a letter to

Davy Stevens. He may be coming by to borrow the tape recorder to listen to it, and maybe record one back to me. And I guess he and I will send that one back and forth.

I got another tape that I'm going to be using here pretty soon that I just sort of stumbled on as I was looking through my locker box. I've found some amazing things in my locker box that I didn't even know I had. But anyway, one night back at Marble Mountain young Tom got drunk, came back to the hootch, and got really funny. He was just sitting up there on one of those wall lockers with one of those g**ner coolie hats on, talking away. Really funny stuff, so I got out the recorder and got about 15 minutes of young Tom in his prime, thoroughly inebriated. I've decided to save it because it's really a classic. In fact it may appear at the first of your next tape. You're welcome to listen to it, but it won't be fit for circulation. It has a bit too much profanity in it.

Well, I can't think of anything else to say right now, so I'll close off and try to talk to you again tomorrow, or maybe the day after.

March 21, 1969

Hi folks, it's John again. The weather here is almost like spring, and the last few days have been interesting, especially yesterday. I wasn't actively involved in any of it, but it happened close by in a little zone called Argon right up in the corner where the DMZ meets the Laotian border. In fact, it's probably no more than half a mile from the Laotian border and almost that close to the southern edge of the DMZ.

Anyway, at one time this piece of real estate belonged to us, so of course we pulled out and abandoned it, the same way we did Khe Sanh and some other places around here. So anyway, our leaders decided to go back in, and do it by surprise. I think I may have mentioned to you in the past that we have certain people in the higher echelons who come up with these creative ideas all the time. Anyway, they tried yesterday to sneak troops into the area without any zone prep of any kind. To make a long story short, the total tally at the end of the day was one F-4 shot down, one OV-10 shot down, two Hueys shot down, one CH-53 shot down, three CH-46s shot down, and four CH-47s, which are Army birds, shot down. That's quite an impressive tally. Also, there were quite a few MedEvacs out there because the guys just dropped into a world of shit. Thank God I wasn't up there involved in it.

Although I was by no means without problems of my own. When I first left the ship going into the beach, I lost an engine. So I had to leave my airplane at Quảng Tri. In fact, I was headed there at the time and went back to the ship with my leader, got another airplane, and went out to do some work. I didn't actually do a whole lot. We were flying C&C missions now, command and control. What it amounts to is you're working as sort of the personal taxi cab for the command of whatever unit you're assigned to for the day.

Oh, by the way, I saw York yesterday. He was with the 9th Marines in Operation Dewey Canyon. That was up at the mouth of the A Sầu Valley. He was at LZ Cunningham when it was overrun. He was actually in Laos doing some work in there, and he said he's never been so scared in his whole life. I can well imagine how he felt. He was telling me that as they pulled out of the LZ down there, the g**ners were coming in as fast as they were going out, and he was one of the last to leave. They were engaged in a firefight even while they boarded the airplanes that took them out. As you can imagine, I was glad to see he got out okay. He still has 38 more days to do as a FAC before he can go back to his Squadron. His orders are in, but of course he hasn't seen them.

I still haven't seen my orders, so I don't know where I'll be going either. But they should be here before long.

I had another interesting thing happen to me yesterday. I came back to the ship and landed, and just as I shut the bird down, the Landing Control Officer came up to me and asked if I was from West Virginia. When I said I was, he said, "You're from Barboursville, aren't you?" So I said, how the shit did you know that? And it turned out he is from Barboursville, too. I couldn't place him at first. His last name is Zoose or something like that. Then I remembered he ran around with Lennie A. and that bunch that used to hang out up at the firehouse. Anyway, he was telling me his parents used to live in Barboursville, but now live in Florida. And after I got to thinking about it, I could remember seeing him around, and maybe talking to him.

Then today I was over at LZ Kiwi which is there at Đông Hà. I had taken the Colonel and the Commodore over there and I was just sort of waiting around for them to come back when this Major walked up to the airplane to see where I was going. Well, I got to talking to him and noticed he was looking at me kind of funny. Then he noticed my class ring and asked if I had gone to Marshall. I said I had, and he said he had once worked around the area. Then I began thinking and darn if I didn't remember him, too. He used to run the I & I staff up there at the Reserve Unit in Charleston. I'd worked with him doing some recruiting down at Marshall. So we had a little chat until my passengers came out and I had to leave. But I thought that was interesting.

I didn't actually do a lot of flying; got in .7 of an hour as a stand-by bird. I didn't get out of the rack until about 11:20. I've had trouble sleeping the last few days, although yesterday I got seven hours in, came back beat to shit, and was out like a log and slept late.

Well, I just hope this type of mission keeps up because it's pretty good duty. It's enjoyable in that we don't get into a lot of bad stuff. My leader took a couple of hits yesterday going into a zone but he didn't even know it until he got back to the ship. I went into that zone too, and it was kind of unusual, right on top of a mountain. The mountain was shaped sort of like an upside-down ice cream cone, and it had a pretty sharp peak up on the top. There was a little bitty wooden platform up there for us to land on. It was so small that when I felt my wheels touch down on it, I couldn't actually see it. In other words, the bird just barely fit on the thing. Zones like that are especially hairy because if you're heavy as you begin to slow down, and you start passing out of translational lift, then the only thing that's keeping you up is the power available. At high altitudes like that, it requires more power to keep the airplane in the air when you're not getting the extra translational lift which increases the efficiency of the blades. We've lost more birds shooting approaches to mountaintop zones like that than we have from any other cause. If you're not careful about how you plan your approach, and you get right in there too close, you won't leave yourself an avenue of escape. In other words, you plan it so that if you begin to lose turns, which is an indication you're not going to have enough power, then you can kick a rudder and slide it off, and you'll have a drop-off that you can use to drop a couple thousand feet. Well, in this case we had maybe 1,500 to 2,000 feet to drop off, enough to exchange altitude for air speed and get our turns back. But I got in okay, picked up the load, and lifted out okay. Didn't have any problem.

Haven't got too much mail of late, probably not since the 15th or 16th of March, which is four or five days. Of course, that's typical for me; I don't get a lot of mail.

Tomorrow I'm going to lead a section out. I'll be working sort of southwest of Khe Sanh out towards the Laotian border. I'm not really excited about it, although I understand the zones are mostly secure there, level zones for the most part. It's like a plateau up there and the guy who flew it today said it was a good mission, so I don't expect any special problem.

I don't know if you can hear that music in the background, but my roommates have been dubbing and copying tapes like crazy lately. In fact, the tape recorders run pretty much full time all day and all night. But I don't mind it. Actually, I enjoy the music.

Well, I can't think of anything else to say, so I'll just close out and talk to you again later.

March 23, 1969

I got your letter today, and Pop was talking about Christmas in July again. Actually, I thought I made it fairly clear I'm not all that hot on that idea. I hope you won't be disappointed, but it really is out of season. Maybe we can just wait until next Christmas, and have a double special one. You know, the Christmas we had over here was actually pretty depressing, but that's all over now, and well….

Anyway, I'm not any closer to Captain, I don't think. I haven't heard anything yet. Well, I've heard they have to promote all of us on the list by the first of July which is the end of the fiscal year, but I don't know yet. Nobody around here seems to know anything. They still haven't gotten down to me on the list.

Also, I want to reassure Pop I have learned to barf and can do it pretty well. And don't worry about me and liquor, because we get along pretty well. I don't let it get in the way of my flying; I can't afford to. Flying with a hangover is just not the thing to do. You can get yourself into a lot of deep, serious trouble that way. By that I mean you can get yourself into situations that you just don't have the edge to handle because your reaction time is so far down. But don't worry about that because I want to get home just as much as you want me to get home.

Also, Pop, you mentioned that you drove the beast for three miles in third gear at 70 which put 4,000 turns on it. That is hardly more than normal cruise as far as the RPMs are concerned. You've got to get it up, well, I think it red lines at about 63 if I'm not mistaken. You've got to get it up there before it does much good. So maybe next time. But I'm sure it helped some.

I was kind of surprised when you all first told me that John F. and Julia D. were getting married. It just didn't seem like a normal match, but to each his own, I guess. I hope everything works out for them.

About the 86 more days flying. I see you are probably keeping better tabs on how many days I have left than I am. But as far as the 86 is concerned, I won't know exactly how it will work out until the last month.

We got a lecture the other night from one of the Majors who said he had heard some negative talk in the Squadron. He said he didn't care if the Army or the Air Force or the Navy or any of the rest of them only stayed in combat six months; that was just tough. We'd signed a contract with the American people and they could within their rights require us to stay on station for the duration. But they were nice enough to ask us only to stay 13 months, and we were going to do our jobs regardless of what anyone else did. So I don't know if that's going to be the policy, but it might be.

We haven't heard anything about Tom's mother, but we do know he is on his way back to Vietnam now. He isn't here yet, though. I imagine he's in Okinawa. They have bottlenecks in Okinawa holding up traffic both ways except for homebound emergency leaves that get priority. So we know he didn't have any problem getting back to the States.

I heard from Bob W.'s wife. His wife lives near Tom's wife in California and are quite close. Anyway, Tom should be back any day now. He's got orders for El Toro with transition to jets which is what he wants. If we were to go home early on account of his mother, he would have to get orders for Washington, which would preclude him transitioning to jets. So he's coming back to Vietnam to finish out his full tour.

When Tom comes back, I'll tell him what you said in your letter. At some point he and his wife will be going east from California, so maybe they will stop and visit you. If not, maybe you can meet him in D.C. sometime. He'll be up there next summer about the time I get home.

About Cathy; I have no idea what she has in mind. I really don't care. I answered her letter, and haven't heard anything more.

"Hi John."

That was Jerry who just came in. I guess you probably heard that loud....

"Hi John. I just wanted to say hi, Mr. and Mrs. Ace, er, Mr. and Mrs. Astle. I don't know why everybody calls you Mr. and Mrs. Ace. It must be because the Ace is your son. I don't know why everyone calls him Ace, either. It's a nickname he thought up for himself, I think."

That's a negative.

"Well, anyway, I just wanted to say hi while John was talking to you. I hope everything is real good at home. Here's John again."

That was young Jerry S. There's a lot of activity in the room tonight. My roommates, as I told you, are working these tape recorders and they work them all day and all night. I've got to get up in the morning about 6 or so for a 6:30 brief. That's about five and a half hours from now, although I did sleep most of the day today.

Well, as I told you last night, I was going to lead a section out this morning. We got a brief in the morning, went to launch out, got up on the deck and from the aft end we could not see the forward end, and vice versa. So we wisely decided to go back and wait. So I went back to bed, of course, but I kept my boots on, and everything. I laid down and went to sleep and slept to almost 11:00, got up and went to chow, and then went to the ready room and played some darts. After a while the weather cleared up a bit and they decided it was time to launch. So without ascertaining what the weather was inland, they launched us. When we got inland, we couldn't get into Quảng Tri because it was IFR. We asked about the possibility of getting special VFR clearance for coming in at low level, something we don't normally like to do. We started in at about 150 feet or so and it got really bad. So we did a 180, came back to the ship, and cancelled it for the day.

So I never did get to fly. I don't know, maybe tomorrow. I got the same thing scheduled tomorrow. I'm not real excited about going because some of the places I'm supposed to be working are only about 1,000 meters from the Laotian border. The locals aren't exactly friendly down in that part of the woods, though it hasn't been as bad there lately as it has in a lot of other places. We won't really be hanging our nuts out that much, so it shouldn't be too much of a problem.

Well, it's 12:40, so I think I'm going to shut this thing off and try to get some sleep. Maybe tomorrow I'll have more exciting stories from Vietnam.

Oh, one more thing before I close down. I got in a big bull session up in the ward room tonight. We got to talking about motorcycles, which reminded me of mine. I've been thinking about my motorcycle quite a bit. I'm looking forward to riding that thing again, and I just want to remind you of my request that you keep it fixed and running. I'll cover the expense. Also, I'll write to Davy Stephens and ask him to help you get it into the motorcycle shop. And I want you to register it in Florida so I won't have to pay that bullshit West Virginia property tax.

About the little card you sent me. The guys thought it was kind of cute, but a man's calling card has to reflect his personality. I think that card was a bit too gaudy for me. For one thing, I want to stick with the idea of having the ace of spades on the back. It's a subtle thing, but the guys know what it means, that it stands for the Ace. You know, when I go into a zone, I hand the card to whoever I'm dealing with and say they've just been serviced by the Ace. I'm trying to get a legend started, so all the grunts will talk about the crazy pilot they call the Ace. Oh well, I don't know.

There's a real legend getting started up here about an Army pilot who calls himself the Judge. Whenever he goes into a zone he'll start his spiral and as soon as he bottoms the collective to start his spiral, he'll key the mic and say, here comes the judge, here comes the judge. So all the guys are talking about the judge now. I've got to get them talking about the Ace.

Well, I'll talk to you later.

March 23, 1969

What a day today has been. I guess I talked to you last night and told you I would be leading a section out there today. I had requested that mission because things had been going pretty slow out there, not much action. That's sort of what I'm looking for, these pu**y missions where I don't have to sweat much.

But today people out there decided to move into a NVA base camp. So I left this morning about 7:30, went out there and hauled people around for a little bit until the Regimental Commander, this full bird Colonel, was ready to go out and do his visual reconnaissance. He wanted to insert some fire bases into the area. A fire base is just a little camp with artillery used to support the grunts working around the section. They usually pick some high ground so you have a commanding view of the territory around there.

This was all in an area south of Khe Sanh down right along the border. In fact, a couple of times today I was actually flying in Laotian airspace. Well, anyway, the Colonel did his VRs and everything and then we took him back. Then we hauled some, but wait, I'm getting ahead of my story.

Another interesting thing happened when we were out there. We saw a bunch of Hueys coming back and they had a rope hanging from the bottom of two of them. Attached to the ropes were people. We heard the chatter on the radio and all of a sudden one of them says, "Hey, I've got to get down quick because one of these guys is about to fall off." So they landed right down there in the middle of the base camp, got all the guys off the ropes and into the birds, and then they lifted off again.

Well, not long after that we saw this big explosion and the next thing we know here are these people doing a MedEvac. We had two 46s down there, but a couple of the emergency ones were in a heavily wooded area, and we were going to have to get some jungle penetrators to get down. Well, none of the other 46s had them. But it happened that my wingman did. So he volunteered his services.

In the meantime I was doing this VR with the bird Colonel on board, plus a Lieutenant Colonel, a couple of Majors, some Captains, and a Lieutenant. Altogether I had nine people.

So my wingman, went over and did some hoists, altogether 18 of them.

So he went down into the second zone, took some fire going in and got as many people out as he could. Then I asked the Colonel if I could go down and get the rest of them out. He discovered they were his people from his regiment so he said, Roger, by all means. So we went down, took some fire going in but no hits, and got the rest of them out. Well, in that particular zone I landed on a stump, punched a hole in the bottom of my bird and broke it. Still, we managed to get out. I took the MedEvacs back to Vandegrift and shut down for lunch.

While I was having lunch, I fell into a conversation with a friend, a guy I went through basic school with, who told me he had gone out with some Army Hueys earlier in the day to do an emergency extraction of a recon team up in a place called Helicopter Valley, which is up near the DMZ. The recon team was in deep shit, completely surrounded by g**ners. The friendlies were running Fixed Wing airstrikes in there and everything trying to get the enemy fire suppressed so they could go in. They finally got down and picked them all up, but the last bird out had one guy hanging on the skids as they lifted out. They were frantically trying to pull him in, but at about 2,000 feet they lost him. So it wasn't all that good an operation.

Well, in the afternoon we went back down there. In the meantime, the grunts on the ground had got themselves into a world of shit, steady contact, and firefights all over the place. They had called in air strikes and Huey gunships were strafing the area, just the usual hell. So we took the Colonel, did another VR, and then I dropped him into this zone. And boy, was it ever bad. I wanted to get a picture of it with my camera, but I couldn't let go the controls long enough. I had the two back wheels resting on the ground and the nose wheel in the air and I was trying to keep from hitting the trees. It was just one great big bomb crater we were in and we just kind of sat there. Boy, was that a bad zone.

So the bird Colonel was going to be there for 20 minutes. Well, I didn't have enough fuel, so I said I would lift out, go back to Vandegrift, refuel and return to pick the Colonel up. He said okay, so we went all the way back to Vandegrift, refueled, and drove all the way back out again. And this area was a way, way out there in bad guy country. Well, when I got back out there I discovered I hadn't burned as much fuel as I thought I would which left me pretty heavy on a hot day, a bad combination. So I decided to dump some fuel to lighten myself up so I wouldn't have any trouble getting into and out of the zone. So I dumped about 800 pounds from my left tank, shut it off, and began dumping from my right tank. But then it wouldn't shut off, just kept on dumping. All I could do was sit there and watch that tank go empty. I got a scare out of that. We had to drive all the way back on what remained in the left tank.

Then we got back to Vandegrift and the crew chief got the problem fixed, or at least got it shut off, so we refueled again and went back out there. When we finally left the area tonight, they were frantically screaming for gunships. One bunch of them were engaged in hand-to-hand combat with the NVA out there. So altogether I flew, well, I was strapped in that seat for 8.6 hours. I'll tell you, it's....you feel it.

I remember Pop used to tell me about working. You know, like the time you had me work down there at the retail store moving those freezer crates around. Well, I thought I was tired when I finished a day of that, but I don't think I've ever been as tired as when I come in from six, seven or eight hours of flying scared. Boy, it just whips the living shit out of you. That's no kidding. Come back and you're just completely drained. Plus, your butt hurts from sitting in that one position all day.

But I got tomorrow off. I don't have to fly at all, so you can bet your sweet bippy I won't stir from my rack until lunch time. It's about 11:30 now, and I have a fierce headache. I think when I shut this recorder off I'll just go to sleep.

Another thing I noticed today is I'm getting shaky because I didn't let my co-pilot fly at all. I'd let go of the controls and I just didn't know what to do with myself and I'd take them back from him. As long as I had the controls in my hands, I felt a little bit better about things. God, was I ever nervous out there today. I came back and I was just knotted up inside; I haven't been like that in a long time. The calm, quiet Ace is, well, it's starting to get to me.

Another problem we're having is this ship. The guys are such assholes to work with. They keep the ship between 25 and 30 miles from the beach which is ridiculous. We came out from the beach tonight and the ship was about 28 miles away. I had to drive out using my TACAN. The visibility was poor and I was about two miles from the ship before I actually caught sight of it. A lot of times the ship's TACAN doesn't even put out a signal, so then what am I supposed to do? I don't know.

Also, the guys on the deck are constantly giving you a ration of shit. I'm afraid one of these days I'm going to get into trouble. These guys....you come down on the deck and immediately they want to know if your airplane is up or down, and they want the answer right now, you know? Well, to give you an example, they didn't mess with me today because when I came back, my airplane was down. There's no question about it and I just told them, yeah, it was down. But this other guy came in and they asked him the same question. He said no, that's a negative. So he lands on the deck and proceeds to check out some of his systems so when he finally shut it down he would know better what was wrong with it and what to write down on the yellow sheet. So they called him up and demanded to know, hey, are you up or down, in a real nasty tone of voices. I mean, they said hey, you can't just sit there on the deck and turn. Are you going to shut it down or are you going to leave? Fortunately, that guy hadn't been flying all day, so he didn't react. But these guys sit out here on this boat and drive around the sea 25 or 30 miles from the coast because they're afraid to get in close. We go in there and hang our nuts out all day long and come back here to the ship and they expect us to just toe the line for them.

So I've decided I'm going on a hate campaign against these swabbies. They've got some full Commanders running the place. They've got too many of them, that's their problem. A Commander is the same as a Lieutenant Colonel. But what I'm going to do from now on is this. The minute I touch down on the deck and I know I'm going to be a shutdown on arrival, I'm turning off my radios. If they want to talk to me, they can wait until I shut down my airplane and climb out. It may piss some people off, but we're going to cure these guys of being so arrogant.

You know, another interesting thing I've noticed and want to get down for the record is how your perspective of time varies depending on your situation. Here recently at night I can sort of shut my eyes and visualize how much time I have left. I'll try to draw you a mental picture of what it looks like. It's like standing on a sidewalk and looking out ahead of you. You recall how sidewalks are marked off into squares. Well, to me each one of those squares represents a month. I can look and see I have about three and one-half left. I think. Let's see, May, June, July, yeah, about three and a half. But you can't jump them all at once; you must take little steps. Each step is a week and I can see about 16 weeks. It really looks good, but I'm getting shaky towards the end, I can tell you.

Well, I think I'm going to close off for now and try to get some sleep. I'll talk to you later.

March 26, 1969

Hi folks. It's 7:30 in the morning, an unusual time for me to be making a tape. But we're heading for....(yawn).... So I thought I'd come down and say a few words to you.

I found out last night that when I get back to the States I have orders to the Second Marine Air Wing, which means I'll be on the east coast, probably at New River. I'm not really excited about it, but I'm not disappointed either because I did prefer the east coast. I don't know what I will be doing yet, but at least I will be doing it on the east coast.

I'm going out flying this morning. I'm not real excited about it. It seems that last night the Colonel, the SLF Commander, found out we had more airplanes available than we normally do. So he suggested we offer some extra ones to the wing in hopes they could find something for us to do. Well, they did, so I'm going out today. I will fly an admin run to Da Nang first. Then I've got to come back to Đông Hà and do some re-supply work. I'll be leading the section today; it shouldn't be too bad.

We have my tape deck out now. You can probably hear the noise in the background. The boys are dubbing tapes like crazy. Using fast speed we can dub a three hour tape in 45 minutes. We've got three tape decks down here all plugged into one another. I've got several good tapes of all kinds of music; classical, rock and roll, mood music, the works. I get these blank tapes for about $2.35 a piece, tapes that cost about $5 retail. And I figure I've got about

$1,000 worth of music already for the price of the tapes, maybe $100. So I should have a good music library by the time I get home. Well, I've got to go to work.

Later.

Well, today has been pretty slow. I went out today and was supposed to run two missions. I got the first one done. I went out to do the second one and I called people up on the radio and told them I was inbound. They said, Roger, are you emergency re-supply birds? And I thought, uh-uh. So I called them back and said, that's a negative. I'm your, and then I gave him the mission number. And then the guy advises me that mission had been postponed until later in the evening. I asked why. He said the company got in contact last night, and that they were still in contact, and just couldn't receive it right now. But they were going to try to get things quieted down so they could receive it later on.

So I said okay. I figured I'd go back to the ship for something to eat. So I started back to the ship and I had to call Đông Hà DASC, the Direct Air Support Center, to check in with them and tell them I was heading back to the ship. So I called in like always and said I was RTBing for a maintenance check. The guy said, say now, we've got this mission and we'd sure appreciate if you'd run it for us. But fortunately, my airplane was down, so I called the guy back and said that was a negative because I've got a downed airplane and I'm RTBing at this time, 6-7, switching channels. And I just switched him right out. So actually I haven't done much at all, but these are the kinds of days I like.

Well, Tom is back now. He got back last night and said it was really good to be back in the States. He sort of hated to come back but felt he had to. In fact, he snuck on an airplane in Okinawa. They were going to keep him there for 30 days, so he snuck out to get on over here.

Well, I'm really looking forward to getting back to the States again. I think about that quite a bit now. I just want to get back and I hope I never have to do anything like this ever again. A lot of times little thoughts come to mind. Like I remember just before I left, Tim G. told me the war was almost over, that they were mostly beaten. Well, he must not have been talking about this particular war, because they're still doing a job on us, especially up here in the I Corps area. I guess you probably heard on the news, although Tom says the news doesn't carry all that much about the war, but it's going pretty hot and heavy here. I hope that things ease up some in the next couple of weeks, and stay quiet for the rest of my time here, anyway.

Well, I think this tape is about run out....

Chapter 19

A Hairy Dilemma

Well, here it is April 4, 1969, and if my count is correct, I've got about 91 or 92 days left. I got my orders. I'm going to New River. I kind of expected that, and it shouldn't be so bad at all. At least I'll be able to make some cross-countries. I've decided one of the first things I want to do is come home and take Pop for a long ride. So Pop, you can be looking forward to that.

Well, I managed to get myself grounded for a week, which isn't bad work at all, if I do say so myself. I banged my knee getting into the cockpit the other day, and I've got some fluid around it now. So I'll be down for about a week, and then after I'm back up for action. It will be only about four more days until we strike out for the Philippines. And then, assuming I get to go with the ship, by the time I get back from that, I'll only have about 60 days left. So that should work out pretty good. I'm doing everything I can to slow myself up, that is, reduce my flying and I've been pretty successful. I think I've only flown twice since the 22nd of March, maybe three times. Not bad, not bad.

Well, by the way, I got a letter from Sabine last night. It was kind of interesting because she hadn't heard anything at all from me since back in February, the early part of February, when I flew out to her hospital to arrange the meeting in Da Nang. She wrote me a letter from An Hoa or from the hospital there as soon as she discovered they were changing her departure date on her, but it didn't get to me until much too late. I wrote a letter to her home in Germany right after I got that message, and now I got this letter from her last night. But what's interesting is that she hasn't gotten home yet, so she doesn't know that my letter is waiting for her there. Anyway, it was a real nice letter. She explained to me what happened, and went on to explain that the airfield was closed at An Hoa. As a matter of fact, I believe they were taking mortars and rockets that day. She said she was so sorry about it, that she had thought all day about me, and how much she would have liked to spend the day with me. She said she really enjoyed the Christmas party, and said she remembered with pleasure my visits to the hospital where we talked while my helicopter was running behind us.

She told me this little story which I thought was rather cute. She said the first time we met she had been waiting at the MAG 16 tower for a ride to An Hoa when the tower operator told her to be careful in the helicopter, that it was being flown by a young, dangerous pilot. Yep. Of course, the tower operator had good reason to think that because when he called down to the Squadron, I was the one that took the call. And he said, well, we have some nurses here to go to An Hoa. And I said, well, are any of them pretty? Yeah, he said one of them was blond, pretty nice. So I said I would be happy to take them out there, but asked him not to mention it to anyone else. Anyway, I guess I gave that operator reason to think I was dangerous. And I am dangerous, after a fashion.

Anyway, in her letter she says she really got to appreciate what a terrible place An Hoa is after she left and went to visit some of the more beautiful places in the world, like in Cambodia and Singapore, and, well, this letter came from Manila in the Philippines. From there she's going on to Taipei, Tokyo, Hong Kong, India, Nepal, Ethiopia, and Lebanon, to see another war she says, and then back to Germany.

She says that when she gets back to Germany she hopes so much to find there a letter from me waiting for her, which she will, and that she hopes someday I will come to Germany to visit her, which I will. One day next July, as a matter of fact. I'm really looking forward to it. After getting this letter, there is no longer any doubt I will make the trip. Of course, you all don't have to publish this all over; this is sort of just between us.

Oh, by the way, being grounded means I have to stay in my rack. This is going to be hard to take, staying in bed all day except to eat. But I think I'll manage okay. I've discovered that the human body can do with a lot more than eight hours sleep. You can push it up to 16 or 18 hours with practice.

Well, I guess I'll sign off. But wait a minute, young Tom as a word.

"Hi folks, this is young Tom back from vacation in sunny California. I figured old Ace was back here on the ship and I just had to come back and see him once again. Thanks a lot for the Easter card, even though one would never know it was Easter here. We're getting so short now we don't even have time to eat all the Easter candy we've gotten. But it's all appreciated, very much.

While I was back in the States, I saw old Steve H. You may remember he was the one talking about the panoply of love versus the panoply of armored plate. He's back at Santa Ana doing real well. His leg is all healed up and everything, and he's started flying again. He's looking forward to coming back over here in about 20 months or so. He's crazy if he does.

This ship is going to the Philippines. Ace isn't going because he hasn't been doing anything for the last three or four months. I think he'll be going to Phu Bai with us even though he has a knee ailment. He ran into a seat trying to get out of an airplane. I guess you told them all about that, though, haven't you Ace?"

Yeah.

"He's already got a flat spot on the back of his head from lying in bed all the time. It's a shame he can't come out and play with the rest of us. He just read us a letter he got from somebody back there in Farbersbakersborough, something about his brother wrecking his car, pulled the carburetor off and leaked gas all over it. He was kind of up in arms about that, but he calmed down eventually.

It's been pretty good out here on the ship. We haven't been doing anything except flying on some C&C missions, taking mortar in the zone and heavy automatic weapons fire occasionally. So it's been real soft sitting around here watching movies. We hope to cruise through the remainder of the tour, just stay out here in the ocean with the squids and kind of cruise around, pop the occasional Darvon, and just take it easy.

Old Ace is going back to New River just like about everybody else in the Squadron. So he'll be back up here with a bunch of freeloaders when he gets back."

What do you mean freeloaders? I've already decided that there's going to be at least seven of my friends there that are married, and I'm going to offer them the opportunity to have me over for dinner at least once a week. With seven dependable friends, that means I'll have a different place to eat every night and will never run out. Then the cry will be, hey, let's all go over to Murph's house on Friday night and get shitfaced.

"Old Murph said Ace is going to be as welcome as a zit on your nose the night of your Senior prom. Ace got that newspaper clipping of his brother with those damn glasses on. Those are the most ridiculous things I've seen in a long time. He apparently was pretty excited about that. Well, that's about all the news from the public affairs office of the ship. I'm going to turn this back over to Ace while I go tinkle. It sure has been a pleasure talking to you again. We'll see you sometime."

By the way folks, I was going to tell you that young Tom will be going to the east coast for a few days when he gets back to the States, and there is the possibility he will come through our part of the country to visit you....

"A slim one."

A slim one, but a possibility, nevertheless. Anyway, he'll be coming along a few days after I do, so you may get a chance to meet him. Also, you probably know I've been living with Norm K., the guy from Rising Sun. But we may get a chance to do some shifting around and get Tom back in here with me and we'll live together again. Yeah, well, I'll sign off now too and talk to you later.

April 5, 1969

Well, we've had another interesting day aboard the U.S.S. Valley Forge somewhere in the South China Sea. We ran into a storm last night, so since about 2 a.m. this ship has been rocking like crazy. Usually, this ship is pretty steady because it's so big and it has a flat bottom, unlike the smaller Tripoli which has a round bottom. But when the seas get heavy enough this boat will rock, and then it rocks even more violently. In fact, the seas got so bad last night that one wave broke over the flight deck which is 60 or 70 feet from the waterline, if you can imagine that. So I've had an enjoyable day going back and forth and up and down.

I also had some exciting news today. I've discovered that the true mission of the Marine Corps is not to suppress the insurgent communist forces here in this county, but rather to absorb all the chicken shit that the heavies can come up with. The Marine Corps in Vietnam is an experimental force so all the heavies in the other services can come up with chicken shit ideas and try them out on us.

Well, it seems that yesterday it was the turn of the guys who run this ship. Well, this boat is a poorly run ship. Let me tell you a little story about something that happened the day before yesterday. They came up with this little message about haircuts. It seems the Captain of this ship has noticed that quite a few of the embarked personnel, and also a few of the sailors, have haircuts that exceed what he considers neat. He also has noticed the beginning of some beards, and also a few moustaches that he does not considers neatly trimmed. So he comes out with this little blurb to the effect that even though social conventions nowadays tend toward longer hair, all hair on this ship will be neatly trimmed. Then he offered up his own little description of what neat is. You know, sideburns can descend down no further than the middle of your ear, and all the rest of that horseshit.

Then he got into moustaches. And neatly trimmed for moustaches means the ends of the moustache must not extend past the corner of the mouth, nor below the corner of the mouth. I'm hoping that this will blow over in a couple of days like the rest of them have, but it came at a rather inopportune time.

You see, the General is coming here tomorrow to pin medals on the guys, and I'm to get my second Purple Heart. Bob W. is getting a Distinguished Flying Cross; we're all quite proud of Bob. And another one of our friends, Ken S., is also getting the DFC. Well, I don't think I will ever get one of the bravery medals. I'm not particularly brave.

But anyway, the Colonel who runs the special landing force has decided that these moustaches are going to have to go, and he mentioned something about it to Lieutenant Colonel T. And of course Lieutenant Colonel T. was forced to mention it in turn to some of the other people. And there you go.

So I've been in hiding. Lieutenant Colonel T. won't get a chance to talk to me today. He won't be back until about 3:00 tomorrow afternoon when the General comes. In fact, he's going to fly the General back, so I probably won't see him until tomorrow. But there is a good possibility that our beloved leader will be forced to require me to trim off the ends of my moustache.

Now up until now Lieutenant Colonel T. had been really good about it, and he's done everything in his power to help me avoid having to trim my moustache. But I don't know how much longer he can continue to fight the system. But I'll tell you one thing. I've decided if they make me cut off my moustache, the Marine Corps will never get another honest day's work out of the Ace. Of course, they're not doing too well now, but I can guarantee they'll be doing even worse. And I can also guarantee them that forthcoming upon my return to the continental United States will be a letter of intent that I mean to resign my commission from this green mother and rejoin the real world.

We have had quite a few discussions about the Marine Corps here, and we agree the Corps is in deep shit. They really have big problems, the big one being retention of junior officers. Since I've been here in Vietnam, the problem has come to a head. There is a lot of discussion in the professional publications, such as Marine Corps Gazette, and a lot of suggestions about how to stem the erosion, but so far no one has yet hit the nail on the head. Last fall we were forced to answer a questionnaire, a study the Marine Corps was doing, to uncover the reasons junior officers were leaving the corps.

Well, the Marine Corps seems to think that the Majority of their junior officers, particularly their junior aviators, are resigning so they can go into the civilian world and make some money. Nothing could be further from the truth. Sure, money is important, but happiness is even more important. I'm sure a lot of them would be happy to stay in the Corps for what they're earning now, I know I would, if working conditions were better.

One thing that really bothers us about the Marine Corps, particularly the Marine Corps here in Vietnam, is that they preferred to purchase things with blood rather than money. Drew Pearson, the guy who writes the syndicated column, recently did an article saying the Marines' equipment wasn't up to par with that of the other services. Well, I can't speak for the grunts, but I can tell you that in Marine aviation, we are using birds that are old and tired. Our supply system is unable to keep up with the demands of the maintenance program. In other words, we go through these birds faster than the Marine Corps can supply replacement parts. As a result, we have very low availability in terms of the number of aircraft available to support the troops in the field.

Our Maintenance Officer, who is typical of the Marine leadership in the higher echelons, continually rides the maintenance people to get the airplanes in the air. So what is happening is they are not being as thorough as they should be, as they have been in the past. So we're having a lot of these little incidents that would never happen if the maintenance people were more thorough.

For example, Bob W. brought an airplane back in the other day and downed it for an electrical fire. Now he didn't know exactly where the fire was, but he could smell it. The maintenance people simply wrote up a report saying they could find no fire, that Bob must have smelled the exhaust or something. The next day, the guy who took that bird out had a genuine fire with 17 people on board and was very fortunate that it wasn't more serious than it was. They managed to land it and everyone got out all right. But it could have been a lot more serious. The problem was that the maintenance people simply had not checked the airplane thoroughly enough to find the problem, which was in the generators. They could have found it, but they're under pressure to get the airplanes up, not solve problems. This is happening more and more. It's getting to the point that a lot of pilots are reluctant to climb up into these pieces of shit. You don't know whether you're going to get shot down, or if the damn thing is just going to fall out of the sky.

Now I don't mean to scare you. These crew chiefs are conscientious, and they do the best they can. But it's still a problem.

Another thing that most of the junior officers are a little bit bitter about is the fact that we come over here as helicopter pilots and have probably the highest exposure rate to enemy fire of any group involved in actual combat here in Vietnam. Even the grunts don't go out day after day. Now it is true our living conditions are pretty plush, and we do skate some. But we expose ourselves more than anyone on a daily basis. We'll come back to the States and we'll be there about two years according to the Marines' master plan. And then there is the chance that those of us who are still in the service, particularly those of us who made the great mistake of accepting regular commissions, will have an opportunity to return to Vietnam.

Now this in itself isn't so bad. But the other thing is that during these two years in the States, if we're at New River, there is a high likelihood we will be required to make a six-month cruise down in the Caribbean. Which means that out of two years in the States, we will probably only have 18 months before we have to ship back over here. Now particularly with young marrieds, this poses a great problem and is a Major source of dissatisfaction with the Marine Corps, as you can probably see why. Now the Marine Corps fails to realize this. They fail to realize that their leadership is not as inspiring as it probably could be. And I've heard some of these heavies argue with these young kids who say the leadership in the Marine Corps is not what they expected. And the heavy says, well, why don't you stay in the Corps and change it. The only problem is that if an individual stays in long enough to get high enough to do anything about it, his mind has been so conditioned by the system that he no longer sees the things that he saw as a junior officer. So I'm afraid I made a giant mistake in accepting a regular commission in the Marine Corps, and I just hope I will be able to get out as soon as I want to.

Another thing, the Marine Corps never takes into consideration the individual case. For example, a friend of ours was flying Hueys over here in Vietnam, and got orders back to the States. He had requested California where his family lives, but he was sent to New River. He is an only child and his father is very sick. He wanted to be nearby to help his family if they needed him, so when his orders came for New River, he wrote to headquarters explaining his dilemma. They answered that if he would be willing to extend for an additional year, perhaps they could work it out. So he wrote them back and said, Fuck You! He is going to get out as soon as he can. Can you blame him?

Or take Bob W. who is a very conscientious officer, or was up until this point. He got screwed because he went to graduate school. He had to wait an additional year to make Captain, a year longer than his contemporaries, because after he was commissioned, he spent a year in graduate school. The Marines ruled that year in graduate school was not as useful as actual Marine experience would have been. Well, that's just wrong. Bob is a very intelligent boy and knows a lot about the Corps.

Now he's working as a Safety Officer here in the Squadron and doing an excellent job. He'd like to go back to the States and continue to work in the safety program, so he requested to be sent to safety school in Monterey, California. He was even willing to extend to get into this school. Well, the Marine Corps not only denied this request, but it isn't even sending him to the Second Marine Air Wing. They're sending him to the Second Marine Division, which means he will be working with grunts. He won't even be in a Squadron. The only flying time he gets will be on his own time. Well, Bob has decided the Marine Corps has got the last bit of work out of him it is ever going to get. When he gets back to the States, he's going to do just enough to keep people off his back, or maybe just enough to avoid getting court martialed. He will patiently wait until his time is up and then get out. You can just see how the Marine Corps wastes its assets.

We've noticed that some of the other services, particularly the Air Force, have very screwed up notions about keeping their pilots happy and letting them fly different kinds of airplanes, and stationing them in places that they request and taking all in all a pretty personal interest in their pilots. As a result, they don't seem to have this retention problem. We feel that eventually the Marine Corps will just destroy itself and no longer exist. It will be incorporated into the Army. It is just not reading the handwriting on the wall.

It could correct these problems, but it just isn't doing it. In the long run, they're hurting themselves.

I don't know why I gave you this little dissertation, except that I'm pissed. You know, overall pissed off. Well, I hope it was enlightening. Of course, I don't want you to go around telling everybody. At least if you do tell everybody, don't tell them where you heard it.

Well, by the way, tell Jim hi for me. And you can tell Ron P. that any time he comes over to talk on the tape, it isn't really necessary for him to identify himself. The last time he talked for a while and then about three quarters of the way through he decided maybe he better tell me who he was. But I already knew.

Next time you see Jack J., I wish you would congratulate him for me on the baby boy. Tell him when I get home I'll take him up on that cigar. Also tell him I enjoyed hearing from him. Anytime he'd like to stop by and talk on one of the tapes, I'll be happy to hear from him.

Things get pretty bad on ship here. Women seem so far away. I don't think I will spend much time roaming around Barboursville when I get back. I don't know that many girls still around I could date. The only need in Barboursville is to see you all and my friends. But then I'll have to get on with my activities, and I'm into chasing women right now. I think I'll probably mosey on up towards the east coast up around D.C. and do most of my chasing up there.

A good friend of mine named Fabo will be leaving Vietnam next month and he's got orders for Quantico. He'll get an apartment up in D.C. and I imagine by the time I get home, he will be pretty well established. And I imagine I'll be seeing a good bit of him.

I talked to my old pal York the other day. He had orders to the Second Marine Division, but I understand they're going to change them for him. More than likely he'll be going to the Second Marine Air Wing, so he and I will probably be living together down in North Carolina. We'll be living off base, I'm sure. Our attitudes are similar. We are going to try to use the Marine Corps to get as much out of them as we can, particularly their airplanes and their gas for cross-countries. Might as well take advantage of a good thing as long as it's available.

I was rather surprised to hear Zeke B.'s voice on your tape. You said it was someone I hadn't seen in a long time, and you were right about that. But you know, strangely enough I knew who it was before you said his name. So if you happen to see him again, tell him I guessed it.

Oh, I got word the other day I will be going with the ship when it goes to the Philippines. Today is the fifth of April, and the ship leaves on the 19th and I won't get back until May 4. By the time it gets back, I'll have about 60 days left. And, I'm hoping I won't have to fly too much more. As you have no doubt noted, I am an accomplished skater, and I will undoubtedly be able to practice my trade when I get back. As a matter of fact, I'm the only guy in the Squadron who does not have a collateral duty. I've managed to skate by. Anytime someone asks me, I say, Yessir, I work out of S-3. One time I actually did work in S-3, and occasionally I do odd jobs for them. But in the meantime, when I'm not flying, I hide in my rack. It's part of my overall program to screw the Marine Corps as they are undoubtedly screwing me.

You know, people talk about all the wonderful deals, but for every good deal there is a price. The Marine Corps gives nothing for free. Sooner or later, they extract their pound of flesh. So my policy is to screw them before they screw me, and it helps my morale somewhat.

By the way, we got word that our former Squadron Commander, who has since returned to the States to work out of Marine headquarters, was stricken with a heart attack. He was put in the hospital and had a second attack. Although we understand he is doing pretty well now, we are all worried about him. I think when I get back I will go visit him. He lives in Fairfax, Virginia, just across the Potomac from Washington. He was probably one of the most outstanding people I ever served under, so I would like to pay my respects.

The other thing I wanted to point out was that when I come roaring through Barboursville on my way to Germany, I would appreciate it if certain family members refrained from advice and counsel. It would be better if they just smiled and said have a good time instead of warning me that I don't really know this girl very well, and that after all she is a foreigner, you know what I mean. I've heard it all. Who knows? I might just surprise you and bring her home. You know, like a story book romance, courtship in war-torn Vietnam, and then I chase her half way around the world. It just kind of gets you right there, doesn't it?

Well, I'm just lying here in my rack smoking a cigarette, talking to you. There's actually not a whole hell of a lot of room in these Navy racks to do all this stuff, but I manage the best I can.

You know, one thing that will seem strange when I get back will be not carrying a gun everywhere I go. I'm going to feel naked without it, particularly in view of all the civil disturbances we've had lately. In fact, I've been considering carrying a gun in my car with me. The way things have been going lately, you never know when you might need one. I'm not adverse to shooting people, so I guess I'll just have to round up all my guns when I get home, and maybe buy a couple more.

I am really looking forward to getting behind the wheel of my car and just ripping around on those mountain roads. I hope you can bring my car for me to drive home from the airport, and I wouldn't even mind getting a ticket on the way home. In fact, I've got it all worked out. The first time I get stopped and asked for my driving license, I'm going to hand over one of my cards. And I've already got the answers all worked out if they ask me where I'm going in such a hurry. I'll just say, well, it's like this. I haven't driven in a long time, and I just didn't realize I was going so fast. Or I could say I was afraid I was going to take some fire and was trying to avoid it, or something. Anyway, I just want to get in it and hear that motor and feel that acceleration.

You just never realize how good life is in the U.S.A. until you've been gone for a long time. As it gets close to coming home time, I find myself getting more and more excited about it.

My knee is better now. In fact, I could probably go fly but since they grounded me for a week, I don't want to disappoint anyone by going to work again. I got my four hours for the month, which is all I need to get my flight pay. So I'm happy.

I got your little card today and it mentioned that my fur goods haven't arrived from Australia yet. I will make a copy of the order sheet and send a letter to them to see if I can't find out what happened to it. I had about $300 Australian dollars' worth of goodies there, and at the rate of one Australian dollar to 86 U.S. cents, that's a tidy piece of change. I'd like to find out what happened to that stuff.

I would like for you to root around in my desk there and find my passport. If memory serves, it expires in May. I would be obliged if you would find out what I have to do to get it renewed as quickly as I can. I won't have any sweat about shots and stuff like that because I've had just about everything imaginable. I have the records all up to date. So please check on that for me.

I was interested in some of the names you pulled off the map there, but unfortunately I've never heard of any of them. If you pick out some a little closer to Da Nang, Phu Bai, Quảng Tri, Đông Hà, places up here, I might recognize some of the names. I can give you some places to look for. A place called Cam Lộ, and another one called Ca Lu. I've flown over those places quite a bit. Of course, you know about Khe Sanh, though I don't think that appears on any maps. We never did get much south of Hội An River. That's the big river that runs just south of Da Nang. You can see it on your maps. It runs east and west and empties into the South China Sea.

When I get home I'm going to try to forget about Vietnam as fast as I can because it's not real pleasant. The other day I was lying here in my rack trying to think of pleasant things, like women, and I had this kind of bad dream that I was out flying and I was in this helicopter. We were just getting the living shit shot out of us and I was watching these bullet holes appear. Then I woke up.

Well, I've run out of things to talk about for now.

April 6, 1969

Well, folks, here's the Ace again. It's been a rather disastrous day here on the Valley Forge. We had the award ceremony today and I got a gold star in lieu of my second Purple Heart. And then later on the skipper gave me the word that my moustache was going to have to go. It seems that the Commodore and the Captain of the ship, who have very little else to do besides run this little world of theirs, have decided that handlebar moustaches are not in keeping with the service. It's so chicken shit I can't believe it. The skipper was in sympathy with me, but unfortunately bigger people than he have decided that's the way it's going to be. I'm pretty pissed about it. I feel as though I'm about to lose an old and trusted friend.

But it won't be lost in vain. I've made arrangements with the skipper to have a Navy photographer come down to the ready room tomorrow and we're going to have a cutting. I will let the skipper do the cutting; it's only fitting. So as much as I had hoped to come home with my handlebar moustache, because of the chicken shit Navy, I won't be able to. And the chicken shit Marine Corps has something to do with it, too.

Well, here is young Tom. Would you like to say a word, young Tom?

"Folks, this has been a truly sad day in the annals of Marine Corps history. The young John Astle has been informed by the skipper of HMM-164 that his close, indeed all of our close friends, must depart. The Ace's 'stache is going to be relieved of its duties watching out for girlies and might unsuspectingly roll down from his nose onto his lip. The skipper came in tonight and said it wasn't his idea, and that's true. But like the Ace said, the chicken shit Navy has caught up to him, and they're going to make him shave the curls off his 'stache. It's going to die tonight. But he's going to keep them and hopefully we'll be able to sew them back on or so after we ride the tide of indignation which has swept the Squadron this evening."

My Squadron mates have risen up in sympathy, but unfortunately we are too few and too small to make any real dent in the Navy, those assholes. This is but another reason why the services are unable to keep junior officers. You'd think that here in a combat zone it wouldn't make any difference whether people had moustaches if they kept them neat, and I always kept mine neat. I waxed it every morning. A lot of times I'd go up and my hair wouldn't even be combed, but my moustache would be neat. So from now on, I get haircuts only when I'm made to go to the barbershop. And I will try to get back at them in other ways, devious ways.

The only reason I really cut it off, or I'm going to get it cut off, is because I don't want to embarrass the skipper. He's really a good shit. If I really forced them to put pressure on me, they would have to put pressure on him first. So I don't want to jeopardize his standing, his career. So I'm going to do it, but it's with a deep sense of regret for the true character of the Ace that will die when the curls of his moustache are cut off. I'm the last holdout in the Squadron. The other guys have already complied, but I resisted until the very bitter end as is my custom. Of course; you and Pop have watched me resist things before, and no doubt will see me resist things in the future. I'm only sorry things had to turn out this way. Had pressure not been put on the skipper, I would have been inclined to fight it myself.

Chapter 20
Romance in the Philippines

Well folks, here it is Monday, April 7, 1969, and we're up here in the ready room for this momentous of all occasions, the clipping of Ace's handlebars. I just thought you'd like to hear some of the remarks that will be made when we actually get around to the cutting ceremonies. In fact, I thought I'd get a few remarks from some of the people standing around and see how they feel about this. Here's J.J., one of my old associates dating back to our days at basic school. J.J.?

"Well, when the word came down that the Ace had to shave off his moustache, he was really shook and just didn't know how to deal with it. He tried all means and ways to avoid it; even appealed to the skipper. He asked me if he could get an admin run into Da Nang and stay overnight so he wouldn't be here when the time came. He even tried to stick the ends of the handlebars into his mouth so it wouldn't show. He actually got through the awards ceremony that way, but he was found out, and now the handlebars have to go.

The word was passed on from the sages of the Squadron that you cannot beat the establishment. Well, they say that Ace tried to beat the establishment by eating it....his moustache, that is. But Ace's 'stache cutting ceremony is about to take place this evening and it's a wondrous occasion for all members of the Squadron to attend. We're getting a big turnout, photographers, the whole bit. It's a momentous occasion in the 164's combat history."

Folks, here we have Major B., one of the heavies in the Squadron, the operations officer. I just thought we'd get an opinion from the other side of the fence here and see how he feels about it. Major B.?

"Well, first of all, I'd like to say that this is a momentous occasion, and I'd like to make a couple of comments about Ace. Major W. used to have my job, and he was known to smoke a cigar or two, I inherited the job from him, and I've also been known to indulge in the occasional cigar, which doesn't bother me too much. In fact, a little chew of tobacco doesn't bother me, either.

But sometimes seeing Ace chew on that moustache is a little upsetting. Then when I went on R&R, Ace inherited the job of S-3, and he tried to uphold the cigar tradition. And the thing that really got him down was that he could do the job okay, but he just couldn't' stand to smoke cigars. We're going to miss his moustache; it's really a sad occasion tonight."

Now folks, we've got before us the Executive Officer, Major Ribbit, Ribbit, Ribbit. I thought maybe held like to say a few words. Major Ribbit?

"Do you have to punch this thing to talk?"

No, it's on now.

"Okay, just one fast comment on this momentous occasion facing the future Governor of West Virginia. I know this moment will go down in history and will be well recorded. It's something he will long remember as he's residing there in the mansion. Thank you, Ace."

Yessir, thank you, Major Ribbit. I have one of my associates here, Major L. I thought he might like to say a few words. Major L.?

"I'm extremely happy to be here at Ace's 'stache cutting ceremony. Actually, the moustache looks kind of like rams' horns as it comes out of his mouth. When they cut the ends off, I fully expect to see blood dripping out of the ends."

Now we're standing here before Major W. one of our more illustrious Majors, who has been with us for many months. He is the backbone of the Squadron; he and I've been together a long time. He watched my moustache grow from its infancy, and might have something profound to say. Major W.?

"Well, Ace, the inevitable has finally come. It's kind of tragic that we have to see this great cultivation of yours finally come to an end. It's a really magnificent handlebar moustache you've grown. However, all good things must come to an end. I'm sure that you'll be happier for it."

Yessir, thank you very much, Major W., for your helpful comment. Now folks, we have the skipper. Lieutenant Colonel T. is going to be the Master of Cutting tonight, and is going to take care of this momentous shearing. And just before we get started, I thought we'd like to get a few pre-game comments from the man himself.

"This is probably one of the most traumatic experiences this Squadron has ever had. It behooves me greatly, as Squadron commander, to be a participant in this ceremony. The cutting of Ace's moustache is by the direction of the commanding officer of the U.S.S. Valley Forge. By direction of the plan of the day, 5 April 1969, all moustaches outside the confines of the mouth will be sheared in accordance with Marine Corps' order and current directives. We hope that when we sever Ace's moustache, he won't lose his courage and personality. It's taken him 10 months to develop this beautiful bush. We hope that he will save it so that when necessary he can paste it back on his lip to give him the necessary ego that's required to be a fearless jungle fighter such as he is. I think we'd probably better proceed on now with the cutting." Folks, I have only one comment to make - remember Samson?

"This is Murph again. The place is crowded with photographers. We've got a pair of genuine surgical scissors and a pair of surgical gloves. Listen to the gloves being put on, folks. (snap, snap) There you are, folks. The gloves are on, and Ace is being blindfolded by his good friend, young Tom.

The Executive Officer, Major Ribbit, will be on his left arm, and Captain Buck will be on his right arm, trying to prevent him from struggling. Here it goes, folks. Here's the left side."

Don't, don't slip skipper.

"I won't son."

"There it goes. Ace, Ace, speak to us."

I'm just speechless at the indignities that a combat pilot has to suffer in the....

"Hello, Mrs. Ace, this is young Tom here at the scene of the greatest sporting event of the year. Young John Astle is standing up there now, the skipper is getting in position. He's leaning over. The air is tense. Two men just passed out. There it is, there's the final snip. Ahhhh, ohhhh. There goes the left side. The left side is gone. Wait a minute. There's an assassin in the area and he's after the skipper. The skipper is over on the outside now. He's getting the scissors into position."

Hold it, hold it.

"He's getting them into position. Here he goes. The scissors are closing up. That's it, that's all she wrote. The Ace's personality just died. The Ace is brushing the final remnants of his 'stache off. He's still got the part over his lip there, but the skipper is getting ready to trim the bottom. He's starting over on the right hand side and moving. He's about to the center now. He's continuing to move up. Ace just punched the skipper. There, he's moving over to the left side, now, getting it all trimmed up nice and even. Ace looks ridiculous. There it is, straight across.

My God, it's not the same guy. The ceremony is complete. Ace is standing there speechless, the final remnants of his 'stache dropping to the ground. Ace is getting some scotch tape ready. He's going to tape the ends so he can save them. The crowd here is just heartbroken. It's like a funeral. Everybody has got their head down crying, tears are flowing. Now the crowd is beginning to get restless.

They're after blood now. Someone's got a razor and is after Kenny S.'s sideburns. Major Ribbit is leading the mob. God, it's a frightening thing to see, a mob rampaging. They're really in a frenzied state.

Well, that's about it from the coliseum here. This has been your friendly reporter, Tom W., signing off on behalf of Fred's Moustache Wax. The Ace is putting the remnants of his moustache in a piece of tape so he can save it. It's all wrapped up. Well, Ace, good buddy, now that the ceremony is over, let's have your words of the hour."

Well, I only feel like half a man now without that lush growth on my upper lip. No longer can I eat things and taste them three days later. It's going to make profound changes in my lifestyle because I'm going to have to do things differently now. And I just don't know if I will have the courage to continue.

"Well, I know what you mean Ace. But I want you to realize that your friends just feel great now that this day has passed. We know how much it meant to you, those of us who have lived with you for 10 months, and watched your moustache grow from infancy to the tremendous, life-giving, probably most vital organ of your body."

I've got one of my old cronies here, and I thought I might get him to say a few words. Here he is, Bob J.

"Well, I'm not a new guy around here by a long shot. I've been here about nine months. And in that whole time, I've known Ace. But just a few moments ago, I met John Astle, the man without a moustache, just another warrior in Vietnam. He no longer has the wonderful moustache he spent 10 months cultivating. The whole Squadron hated to see that moustache go. It was a status symbol, not only for him, but for every member of the Squadron."

Folks, I've got another guy here, another one of my old compatriots, the Incredible Hulk. The folks back home call him Lash. Lash?

"All right, John. I know this is a sad day for you. What can I say? A tear just comes to my eye when I think about what they've done to you. I can't go on."

Well, folks, the Ace is gone.

April 8, 1969

Hi folks. Today is Tuesday, and it's been pretty slow for me, although others in the Squadron have had some excitement. Take my roommate, for instance, young Tom. Young Tom left this morning, or was getting ready to leave, and decided to take a head call before he left. So he did. He went up and made a little head call. Got in the airplane and had no sooner gotten started than he knew he had to, had to....

"Had to take a number-two. Yeah, an exploding number two."

Tom, would you like to tell this, just to make sure we get it straight, you know, for posterity's sake.

"Hi folks. Well, this is a little embarrassing. But seeing as you're Ace's parents, and part of our extended family, I suppose I can go through with it.

Well, as Ace was saying, I was up there turning on the flight deck and I just knew that this number two was coming because I'd had this greasy cheeseburger last night, and greasy cheeseburgers do it every time. But I had already made this head call, see, and I figure maybe I can last anyway until I get to Vandegrift, which is probably about 60 miles from where the ship was this morning. So I got up there and decided, well, we'll give her a try. So we launched and no sooner did I get in the air than I started getting the, well, you know how it is. You get all tensed up, your mind starts to wander. You get all crunched over in a little ball.

Well, Vandegrift is about 15 miles from the coast. So we got into Quảng Tri, which is right on the coast and boy, the pain was just something else. I just had to get down somewhere and take my number two. I decided the best thing was to make it into Vandegrift, which was only about 15 miles away.

So I went in about as fast as the airplane would go, we sort of sneaked in there, and parked the airplane. And boy, it was just backed up something awful, at least 80 pounds per square inch. My safety valve was just about ready to go. But I hopped out of the airplane and said, well, if I run, I might be able to make it to the head across the LSA which was about a 50-yard sprint.

Well, you guessed it. My safety valve broke and I just shit all over my skivvies. I got over there by the recon pad, got in that one-hole thing out there with the screen door on it, and all these recon guys were sitting around watching me wipe my skivvies. It just exploded. It's just one of those things, you know. It just catches up to you sometimes. So there I was trying to clean my skivvies out and make the best of the situation under the circumstances. And those guys, well, they didn't look too favorably on me.

Well, I came back out. Luckily, we had to come back to the ship anyway, so I was able to discard my skivvies which were no longer in serviceable condition. Now I'm down to three pairs of skivvies. So it looks like it's going to be 88 long days before I ship out. I figure three into 88 goes about, oh, maybe 29 times. So I figure I'm either going to have to wear one pair of skivvies for 29 days, or else wear them every third day 29 times."

If you wore them every third day, you'd never be able to get any of them in the laundry, because the laundry only runs once every eight days.

Well folks, as you can tell, young Tom had a hard day of it. That happens sometimes, the trials and tribulations of a combat pilot. Well, I notice this side of the tape is almost gone.

I found out for certain today I am going to the Philippines. That's going to be a good trip. Yeah, that will be a real fine trip. I'm looking forward to it. I told the guys today that if they wanted me to buy anything for them, I'd do it.

Oh, by the way, I invited anybody in the Squadron today that was going from the west coast to the east coast to stop on by. I told them I'd give them the address. But I don't know if any of them will take me up on it or not. But you might want to be prepared, just in case. They're a pretty good bunch of guys, for the most part. No, I don't think they would gross you out too much.

You'd probably enjoy their company, most of them.

Well, that's all for now.

April 10, 1969

Well, folks, I'm really beat to shit tonight. Went out today, flew up north. First time I've flown in about a week or two. Well, the first time I'd been up north since the 22nd of March. Tom and I flew together today, in the same airplane. Really enjoyed that. He and I get along in the cockpit real well, make a pretty good team.

But the mission today was a C&C mission, just doing little diddly things for people. Like we spent about an hour or so working around this one fire base bringing stuff from the upper zone to the lower zone. And hell, they weren't more than 100 meters apart. But they take these airplanes up there. It's costing the government about $160 an hour to run them; a pretty high-priced trucking service.

In one of the zones today I came closer to crashing than I have since I've been in Vietnam. It was all my fault. I just screwed up the approach. I was tired and misjudged it. I didn't see my mistake until I was down pretty close and then the whole world just turned to shit. What happened was I was shooting an approach to a ridge line and I misjudged my altitude, my rate of descent. So when I came into the zone, I was a little bit too high. I should have waved it off and started over again, but I tried to salvage it. I thought I'd get over the zone. It's just a ridge dropped off on both sides. It was kind of narrow on top. We were making two wheel hover landings and things like that up there. But I figured I get over the ridge and I'd be in ground effect and I wouldn't have any problem.

But when I got in over the zone, I was still too high and started losing turns. I don't know if I ever explained to you about turns. Turns are the lifeblood of the helicopter. That's what keeps it flying. The turns we speak of are the RPM of the blades. And in this particular airplane we measured that 100 percent was normal operating range, and zero is dead stop. Well, I went down to .78 percent today and the airplane just sort of wobbled out of the sky. We sort of came into a controlled crash into the zone. I was pretty lucky. The only right decision I made was not to try to wave it off after I started losing my turns. I just went ahead and drove it on into the ground.

We were lucky. Had we rolled or had I tried to wave that off, we'd have probably just rolled that thing up in a little bitty ball because the zone was about 2,300 feet on top of this ridge line, and it was probably a good 1,000 feet to the other side, to the floor of the valley on the other side. It's kind of exciting. I'll tell you, you go out and do that for, well, we were out for 7.8 hours today. And as I've mentioned many times earlier on these tapes, it just literally wears you down to nothing, particularly if you're scared part of the time. That really does it to you.

I was kind of ashamed of that approach. I'd never done anything that bad before, but I was tired and I just, well, I guess some days you have the touch and some days you don't. Today I'm sure I didn't have it. It was probably a good thing Tom was with me.

Well, one other little story. We pissed off the General today. He sent his errand boy out, a full bird Colonel. He came out to tell us to get our airplane out of that zone and never come back in there with a 46. We just kind of grinned at him and lifted out. Those dumb shits. All they have to worry about is whether or not 46s land in the General's zone. Assholes. They ought to have to go out and hang their nuts out flying around here sometime.

Another guy today was bad mouthing our Squadron. These guys were pissed because we weren't doing things exactly the way they wanted them. Went up and we're moving these things from the upper zone into the lower zone. The first load we took in, the guys were pissing and moaning because we didn't drop it exactly where they wanted it. They popped smoke for us and we put it right down on the smoke. Apparently, that wasn't where they wanted it. We were having trouble with our radios and couldn't talk to them. Finally, we heard one of the guys say that our Squadron was a pretty sorry bunch. So old Tom, he just went ripshit. We got into the zone the next time and he sent the crew chief out to grab that radio operator by the throat and drag him back to the airplane. We took him with us to show him because they were pissing and moaning. Like one time, we were only able to pick up four people. If you can imagine. You've seen a 46 and you know how big it is. If you can imagine that four people was all that we could pick up. But any more than that and we never would have gotten the thing off the ground.

So we dragged this radio operator off and Tom brought him up in the cockpit. He was just chewing him out something fierce about making negative comments about our Squadron. Then he pointed to the rotor tach and said, you see that gauge? He said that when that gets down below 90, this airplane does not fly. And we were carrying about 96 percent right then because we were heavy. And it was hot outside today and the zones were pretty high. And it wasn't long after that I crashed into that zone.

I think Tom's performance kind of scared the radio operator. Tom explained to him the difference between the Alpha and Delta models. You see, we've got the old Alpha models, and they're underpowered. Delta models could go up there and pick up 10 or 15 people in a 2,500 foot or 3,000 foot zone where we could not. We were going back to pick up some externals for them until we got up there and discovered that the damned things weighed almost 2,000 pounds. There was no way we could pick them up. We tried, but we couldn't get it off the ground. You start pulling in power and your turns just bleed off.

Finally, we just put the radio operator back on the ground. In this particular zone, you had to make one-wheel hover landings. In other words, you get over the zone in a hover and you back into it and you rest one wheel on the ground and lower the ramp. Your other main mount and your nose gear are in the air. You're holding it in a hover. That's a pretty tricky one to do.

Well, I can't think of anything else to say tonight, so I'll close off.

Sometime later....

Well, folks, I just got back from the Philippines where we had a brief respite from the war, and I can assure you we made the most of it. Now, actually there isn't a whole lot to do in the Philippines in terms of what we would call culture. But when a bunch of Marines wade ashore after leaving a hot war zone, they have a way of making things happen on their own terms.

Well, that would more or less describe our experience in the Philippines, leastways what I can remember of it. I distinctly remember getting shore leave and heading off with some of my pals to do a little bar hopping. After that, things get a little fuzzy.

I remember we were in the officers' club there tossing back a few drinks, the way boys will do, when they told us it was closing time. Well, we took that news in a negative way because we were by no means ready to quit. I mean we've been tooling around out here in the South China Sea for months, getting our ass shot off on a regular basis, and we finally get a chance to relax and let off some steam. And this guy comes around and tells us it's closing time. Now what were we supposed to make of that?

Well, anyway, it turns out that one of the boys knew about this native club out in the ville away from the post, the kind of place that caters to military personnel like ourselves after hours. So we all pile in a couple of cars, I can't rightly remember where we got the cars, and we head off for the ville. And after driving around in the jungle a while we finally find this place which is tucked back against this mountain on the side of a hill. In fact, the parking was built on a steep hill, which is something you should keep in mind because it figures in the story later on.

So by the time we got into this club, we were pretty well along, as you can imagine, and this place was really rocking. I mean, they had a native rock band going full blast, and a wall of military people in various stages of inebriation having a good time. The lighting wasn't real spectacular, and it took us a while to get adjusted, but we finally found a table and ordered a round of drinks. By that time, we were pretty thirsty, as you can imagine. And after a bit, when we'd had a round or two, I chanced to spy a couple of lovely young damsels sitting alone a few tables away from us. With minimal encouragement from the boys, I just sort of sauntered over and invited them to join us. And me being a handsome, devil may care pilot and all, why naturally they were charmed and immediately got up and came over to our table, and sat down.

Now probably here I should probably inject a note that there was a reason these girls were alone until we came along, and that would probably have something to do with their girth. The first one, and the one that took my fancy right off, was named Pamela, and I guess she weighed in about, oh, maybe 190. She was a bit taller than me, maybe five foot ten and every inch a beauty. I couldn't take my eyes off her.

The other girl was a lot bigger. I never did get her last name, so we just called her Helen the Bruiser. You could just tell right away that old Helen was not to be trifled with. She had arms like tree trunks, and a tattoo on her right bicep. It was a snake of some kind winding around a pole. It was actually quite attractive, but Helen seemed a bit standoffish. I suppose she was concerned about protecting her virtue, and who could blame her? Certainly not me. A girl can't be too careful these days. Also, the maidens were the daughters of officers, which can make a girl defensive.

But anyway Pamela and I hit it right off from the start so much so that I was afraid one of the other guys might horn me out. You know, you can never take anything for granted when a pretty woman and a bunch of Marines are concerned. So after a bit, when the band finally started playing a slow song, I invited Pamela to dance, to which she graciously consented. I tell you, my heart skipped a beat as I led this beauty to the dance floor. And all my friends were yelling and cheering for me too, which sort of made me swell up with pride. We Marines are always eager for approval from our fellow Marines.

So Pamela and I are out there dancing away, Just in seventh heaven, having a wonderful time, when I accidentally stepped on the foot of this big sailor who was sitting right next to the dance floor. Well, this guy really took offense at that and really gave me a skinning. I would be embarrassed to tell you some of the things this fellow called me, and he didn't even know me. Suffice it to say he called me just about everything but a brave Marine from West Virginia. I would have defended my honor except that this guy was about, well, let's put it this way, he was bigger than Pamela. In fact, he was almost as big as Helen. Well, there was no way I was going to give this guy an opportunity to work me over. But I was thinking, you know, considering all my alternatives, like us combat officers are trained to do. And I thought to myself, okay buddy, if that's the way you want to be.

So I didn't say anything and kept dancing with Pamela, but after a bit I sort of got her turned around and moving back towards that sailor. And well, you know, with a woman that size, when you get her started in one direction, it's kind of hard to get them turned around. They are sort of like ships in that sense. It just occurred to me that maybe that is why ships are always called she.

At any rate, I had Pamela going at a pretty clip when she bore down on the sailor who, unfortunately for him, was looking the other way and didn't see her coming. And just about the time he sensed something was up and looked our way, it was too late. Old Pamela came crashing over him like a tidal wave on the beach. She fell into that poor sucker's lap, just like that. You couldn't even see him underneath Pamela. All you could hear was a long hissing sound as all the air escaped from his lungs. And then there was this awful creaking noise as the legs of his chair gave way, and he and Pamela sort of dissolved into the floor together.

Well, of course, Pamela was just as apologetic as she could be. And the sailor, well, what could he say to this young woman who had just accidentally fallen into his lap. Well, nothing of course. Because if he did, Pamela would have knocked the shit out of him. And if Pamela couldn't do it, by God Helen could.

Anyway, after this, all my pals were cheering and making the greatest noise you ever heard. And as you can guess, I was just as proud of Pamela as I could be. In fact, I decided then and there that she was the woman for me. I forgot all about Sabine and the girls in Barboursville. I could see no other woman than Pamela. In fact, I couldn't see anything but Pamela. She had a way of filling up the horizon.

Well, we went back to the table and everybody was just all over us. The guys made out as how that was the best performance they had ever seen, and they kept buying us drinks. And after a bit, I must confess, I think Pamela and I were getting pretty drunk. Well, actually I was drunk before I got there. And then I had some more to drink, and I can drink pretty fast sometimes when someone else is buying the drinks, as was the case that night. I mean, that's one thing about Marines. You put on a good show for them, and they will buy you drinks.

Anyway, after a while I decided that maybe I would put some moves on old Pamela. You know, strike while the iron is hot, if you get my drift. So I said, mmmmm, well you know, it's pretty hot and stuffy in here, why don't we step outside for some air.

Well, Pamela avowed as how that was a pretty good idea, so we stepped outside in the parking lot. And then I said something about how it was a little too cool for comfort, so why didn't we step over and sit in the car. And she avowed as how that seemed like a good idea, too. So I led her over to where our car was parked, and gallantly opened the door for her.

Now here you may recall something I said earlier about how the car was parked on a hill. Well, I was a little under the weather by this point and I made a mistake. Now, being a Marine pilot with combat experience, I know what it's like to go into dangerous situations. And in that kind of situation, it is imperative that you keep your cool and watch what you're doing. And so I ought to have known not to get in the back seat of that little car on the downhill side of a fat woman.

But that is what I did. We squeezed into the back seat with me on the downhill side, and I immediately tried to put my arms around Pamela. Well, of course, they would only go part way. Actually, it was about all I could do to reach both of her armpits. If you can visualize a medium sized pair of pliers trying to get a grip on a soda can, that's about what it was like.

At any rate, as I was trying to hug Pamela and get a smooch, she sort of tipped towards me, you know, in a receptive kind of reaction, and in an instant I could feel that a decisive loss of balance had occurred. Before I could react, Pamela was bearing down on me, forcing me back further and further until I was on the floor, pinned in between the rear seat and the back of the front seat. Oh, you don't know what a helpless feeling that is. Then it was my turn to feel all the air rush out of my lungs. It was only with the greatest effort that I could inhale, and each breath seemed harder than the one before. I appealed to Pamela to push herself up, and she really did try. But she was about as drunk as I was by that time, and simply couldn't move. She weighed too much.

So we just lay there as helpless as could be, as I felt my life slipping away. I know it may sound funny now, but it wasn't funny then. I literally could not breathe; I was being suffocated. I had visions of what my obituary would read like, having escaped so many times in jungle combat only to be squashed to death by a fat woman outside a rural bar somewhere in the Philippines. I was wondering how that would read on my war record.

But after a while, just about the time I gave up all hope, the guys came looking for me and found us. Of course, as you can imagine, it took them a while to stop laughing. But when they did, they helped get Pamela off me, and one of them gave me artificial respiration to get me breathing again.

I'm not sure what happened after that. I think we went to another bar, but I couldn't swear to it. But I'm sure whatever it was, it was honorable and fully in accord with Corps tradition.

At any rate, here I am back on the boat safely back at war again. I am happy to report that the curls on my moustache are starting to grow back again, and I'm able to get a little wax on them now. So I'm going to try to sneak by and see if I can have my handlebar back by the time I ship out of here.

I just got your tape, and I wanted to respond to a few things. First of all, Jerry S. is no longer with us. He was transferred to another Squadron. As I understand it, he has gone to the First Marine Air Wing and has turned in his wings. It seems the first day they arrived at Quảng Tri to check into the new Squadron, they lost two airplanes and crews. That made four in two days, so I don't blame him. I think I would give some serious consideration at turning my wings in if I was confronted with a situation like that just right off the bat.

Well, the Marines were kind of disappointed by that. They had screwed up and come up with a pilot shortage in one of the Quảng Tri Squadrons, the HMM-161. It seems that last year about this time the HMM-161 departed from the States as a Squadron. They were at New River when I was there, and then they departed with airplanes and people. So this meant that an entire Squadron was going home at the same time. The people who plan these things just do a terrible job. They didn't mix the crews up enough and now they don't have enough people.

We had a mix master last fall where we tried to prevent this from happening, but they didn't quite get the job done. So we had another mix master. So Jerry and five other people were transferred out to this northern Squadron, and then we got six new co-pilots to replace them.

If you see Bob A., tell him I enjoyed hearing from him, and that when I get home I will come around to discuss my taxes. I will be filing late this year, but I understand we have a 90-day grace period after we get back to the States. So that shouldn't be any problem. Also, tell him not to feel guilty about not being over here; tell him to feel good about not being over here.

About the pictures you asked about. The one with me and another guy sitting in the latrine; the other guy is Tom. The second picture was taken in the coffee mess, and the other guy was Lance Corporal Johnny C., the enlisted guy who worked for me when I ran that coffee mess.

You mentioned the idea of having an open house when I get back, and that sounds fine to me. Anything you want to do is fine with me. I just don't want you to go to all the trouble of putting up a Christmas tree and all that. I am getting short now, real short. I have what we call a short-timers calendar on my wall with various pictures to represent the target day of going home. We mark off the day's one at a time in a daily ritual, and we keep close track of time. And by the close of today, I will have 58 days left.

I got some stuff for Grandma and Betty while I was in the Philippines, some wood carvings she might like to decorate her house. I actually went to the factory where they carve these things and was able to watch them work. It was interesting, and I bought a whole bunch of stuff.

I got about $460 worth of stereo gear, just about everything I need to complete my set. I've got it all, and have it all set up now in my room except for the turntable. I don't have any records, so I'll just leave that boxed up. But I have the tape deck and the amplifier and the two speakers all set up. And of course every night we rock the ship with music. I have these huge, tremendous speakers that really put out some sound. I'm going to enjoy getting back home and getting these things all set up.

Next time you see Ron P., tell him we really enjoy his jokes. And tell Jim T. I enjoyed hearing his voice again. I'm expecting another tape from him, and also a tape from Davy S.

Well, a few words to update you on what we're doing. We're back on station now, the Squadron is aboard, the battalion is aboard, and in the next day or so we will be starting another operation. Things have been getting pretty hot in the Da Nang area, and also up around Quảng Tri. I'm just as happy we're not flying up there. They're losing a lot of airplanes up north, and I think I may have mentioned to you that I've had enough of that sort of thing.

We're going down south of Chu Lai again, south of where we had the last operation. I don't know exactly where yet, or what either, but it appears we will be doing company-sized landings. We'll pick out an area, drop in a company, and let them thrash around a while. If they run into trouble, we'll be standing by to back them up with the rest of the battalion.

These guys have been in the field almost continuously since December, so long I can't remember when they first went out. We pulled them out in the first part of January, right out of the field. They had maybe a day or two aboard ship, and we sailed south and put them back ashore. They were just cut to ribbons down there, really took it in the ear, a lot of casualties. We pulled them back out of there.

Then they had one day on the ship and we put them back in the field southwest of Da Nang, and they've been out there since February. It hasn't been so good for them out there, so now we have them back aboard ship. They can clean up and relax a little. I think that's a good idea, because those grunts really have it worse than anyone else.

Today I'm flying the admin run, which is a good deal, no sweat. I've got a Lieutenant Commander as a co-pilot. That's the same thing as a Major in the other services. But he's a pretty good guy, and we had a good time out there this morning. It was kind of an odd situation because I was a First Lieutenant, and the Aircraft Commander, while this guy who outranks me was flying as co-pilot. The aircraft is my responsibility even though he outranks me. But I've flown with him several times and we get along real well. He's a pretty good pilot.

I was interested to hear confirmation of the Gigi rumor. All I can say is sorry I wasn't in there. I didn't know that was going on, or I would have been over there trying, I guess. But I hope everything works out for her. She's kind of a good kid, and it's just a shame she got in a situation like that. I hope if Tim is the man, he is man enough to stand by her. I also hope that the people in town don't give her, you know, all that much trouble. These are times when people really need their friends.

One of my buddies has a sister who is pregnant now, and she is having quite a bit of trouble because her family has all but disowned her. We were discussing it, and I commented that this is a time when she needs her family more than ever. It's really a shame that they've backed off and not given her support. I think her brother is going to stand behind her, which is good.

Another side note along these lines, we have a young guy here in the Squadron who just found out his wife has taken off with a hippie. He left yesterday on emergency leave to try to get things straightened out, and I really hope he does. He's a really good guy. You know, the women don't realize it, but indirectly they cause a lot of deaths this way.

"Hey John, this is Hog. You've got a 1330 launch. Probably a 1300 brief with Lieutenant Commander Brown, and Heavy Flying is your co-pilot."

Folks that was Hog D. who just came down to tell me I have to go fly. So I have to eat and make a head call. It's 12:30 now and I'm running a little late. So I'll talk to you later.

Chapter 21
Counting Down

Hi folks. Today is, or at least it was, the 6th of May, 1969, and what a day it has been. I don't want to talk too long tonight because Tom's asleep and I don't want to wake him up. It's after midnight now.

I launched out this afternoon about three o'clock leading a section on an admin run. I ended up getting back in tonight about 10 minutes after eight, and boy was it ever dark. I'll just give you a quick rundown on what happened today, and then I'll try to talk some more tomorrow. I don't think I will be flying tomorrow, so I should have some time.

When I launched out this afternoon, I had on board two engines in sealed containers that weighted about 2,300 pounds. Also, I had onboard a couple of passengers and about 1,800 pounds of fuel, and it was 32 degrees centigrade. According to the charts, I could carry about 2,600 pounds.

So I launched out. I drooped just a little bit getting off the ship, but I got airborne with no sweat. I climbed up about 1,000 feet and the wingman lifted off and joined up with me. We started toward Marble Mountain and Da Nang harbor. Got right out in the middle of the harbor and all of a sudden, without any warning whatsoever, my number one engine dropped off the line.

At 32 degrees centigrade, with 2,300 pounds onboard and 1,800 pounds of fuel, a CH-46 will not stay airborne. But anyway, we got all set up for a single engine at 70 knots of airspeed, about 94 percent of the rotor tach, but I still couldn't hold my altitude, and I began descending.

My co-pilot hit the fuel jettison and started jettisoning fuel on both tanks, which you don't normally do. Normally you hit one and then the other, but this time we had no alternative. We had to get rid of weight, and we had to get rid of it in a hurry. In the meantime, he's trying to get a relight on the engine, trying to get it started. We weren't having any luck. Every time we'd get it started, it'd drop right back off the line again. In the meantime, we kept losing altitude. Finally we ended up about 10 feet off the water at 70 knots, about 92 percent. At one point, I drooped down to about 88 percent trying to get my rate of descent up. When I finally got it almost stabilized, I realized I was still descending slowly and there was no way I would make it to land. It looked like I was going to have to put it down in the water.

Well, I wasn't really keen on that idea. I wasn't real sure I'd be able to get out of that hummer once I got down in the water, and I would sure as shit hate to drown in Da Nang harbor. So at the last minute, my co-pilot managed to get my number one engine started again and it stayed started. So immediately we began to climb, getting some altitude, because altitude is your best friend in this business.

Well, we headed in toward Marble Mountain and called them up to declare an emergency, told them we were coming in on a single engine. We were running against the traffic pattern, so they had to clear a traffic pattern for us. Actually, at the moment I called Marble I wasn't single engine, but, I didn't want to take a chance. And it was a good thing I didn't because no sooner had I called them and told them I was single engine than that son of a bitch dropped off the line again. But by then it was no sweat, we were high enough. So I just drove it right into Marble and made a single engine landing.

I left that airplane and my wingman flew me back to the ship to get another one. We hadn't had that one out for 20 minutes until it went down and we had to bring it back. Got another airplane, went out, finished the work, and came back. By then it was, oh, about 6:30. And then those assholes launched me out again. It was after official sunset, and they launched me out to take one passenger to another ship. It wasn't even a code, a Colonel or above, just one measly old First Lieutenant.

So when they called me and told me about that, I said, well, give me a position. They said, ah, wait, one. So I waited one and the guy comes back and says, Roger, the position is one-six-zero. And it shut off.

So I called him back and said, one-six-zero at what? He says, ah, wait, one. He came back and says, Roger, it's about one-six-zero at 10 miles. I said okay and launched out. Well, at one-six-zero at 17 miles I called him up and said, listen, I'm at one-six-zero at 17 miles and I don't see a ship. The guy says, Roger, I suggest you enter a right square search pattern.

I thought to myself, are you shitting me? It was dark by then, and we could hardly see anything. So we've driving around, fiddle-farting around, and couldn't get a TACAN lock on the ship. We couldn't get an ADF lock on. We finally established radio contact, but we couldn't get our UHF homer to lock on. Finally, the ship gave me their position from Da Nang. From the Da Nang TACAN they gave me the bearing and distance, or radial and distance. So I managed to find them. I made a night landing, dropped the passenger off, and launched out. I flew back about 30 miles over the water in pitch black, couldn't see a thing. It was all instrument flying. Got back to the ship and by this time it was after 8 o'clock. I missed chow. I'd missed two sittings. They have two sittings and we managed to miss both of them.

Anyway, we started back into the ship and I know you've never done this, but it's one of the hairiest things I've ever done, that one and a couple of others. But all you can see of the ship are the red lights that mark the deck area. You can't see a superstructure. You can't see the hull. All you can see are these two rows of red lights floating in the blackness. You can't see the ground. You have no idea, I mean relative distances are just non-existent. So I came in astern from the ship and I started down and all of a sudden I realized I was too high. Then I started getting vertigo. So I had the co-pilot get on the controls with me, turned away from the ship and did a 360, a descending 360, which is not a real good idea, but I had to lose some altitude to get down there. I finally ended up doing a hover air-taxi for about the last 200 yards after the ship. Boy, was I ever glad to get that son of a bitch onto the deck.

So, all in all, I guess you could say it was a pretty taxing day. First off, the engine failure took a lot out of me because it took every bit of skill I could muster to keep that thing in the sky. I was kind of proud of the job I did when I got it down, saving the airplane and all the cargo. But God, did that ever take it out of me. Actually, I think I aged about 10 years. In fact, when I got back, one of the guys commented that I looked a lot older than when I went out. Boy, that's for sure. At least I felt a lot older. And the flying out there at night in the South China Sea looking for a ship is not great fun either.

Well, I think I'll call it quits and go to bed. But I did want to mark May 6th on your calendar as a black day for the Ace.

May 11, 1969

Hi folks. Today is the first day I've flown since May 6 or thereabouts and once again it was a black day for the Ace. For that matter, it was also a black day for young Tom. Right after sunset they decided on the spur of the moment to launch three airplanes to go to another ship and bring a company back to the Valley Forge. Okay, that in itself isn't such a bad job unless you try to do it at night. Then it gets really hairy.

Now Tom had the MedEvac bird tonight so he was already doing something when they diverted him to this thing. They launched me in another bird. We started out and the other bird went down and so we had to come right back, shut down, and wait for them to find him, which they never did. We'll they finally found him in the airplane later, but that's another story. I'll get to that presently.

I went down to the ready room, got my aircraft assignment, went up to the line shack to sign it off, and discovered that somebody else was flying that airplane at that particular moment. So I called back to the ready room and told them the airplane they had just given me was being flown, so couldn't they perhaps find me another one. Okay, Roger. They whipped out another number. So finally I got another airplane, launched out, and went over to this little bitty ship. Well, it's hard enough landing on these little bitty ships in the dark where you've got just a little bitty square pad on the aft right there on the fantail. All you have are three little lights, that's all you can see. They have no visual references to judge your closure rate, or your relative motion, or any of that. It's difficult to judge your altitude. It's difficult to judge your distance from the ship. It's difficult to judge everything. In fact, young Tom came close to killing himself tonight.

So we got notice to get three loads out. I carried two and Tom carried one. Then I came back from my second load to the Valley Forge and they told me I was going to be diverted to another mission. I was to take some Colonel into Da Nang.

Well, you see, the thing that got all this started was we were about to go into this operation tomorrow, and sometime tonight they decided we needed this other company over here. Now they had all day to do it, but they didn't get around to deciding until after dark. Unfortunately, the guys who do the deciding aren't the same ones who do the flying. I'd like to put some of those guys in a cockpit out here in the dark and let them see how much fun it is.

Anyway, I took the Colonel into Da Nang, sat down and waited for him about an hour, and then brought him back out here to the ship and shut down. It was after nine o'clock and pitch black. It had been pitch black ever since the sun went down. But the real clincher was that later on they decided it was all a big mistake and that we should just take all these people back to where we had picked them up. I thought, are you shitting me? We went out there and hung our nuts out because we were told it was important to get those people aboard the Valley Forge, and then all of a sudden they change their minds and say we have to hang our nuts out again to take them back.

Well, young Tom, who has his shit together, asked them if it was an emergency, and they said no. So he said he was shutting down, and so were the other two birds. So they all shut down and the Marines couldn't go back. I guess they'll have to spend the night on board the Valley and go back tomorrow. But that, folks, is the Marine Corps and the Navy that are running this portion of the war. I'm really amazed they managed to get this far. But believe me, flying over the water at night is hairy, wouldn't you say so, Tom?

"Yeah, I'd say."

Tom is pretty busy now. He's writing a letter to his wife, a serious letter.

Well, I've only got 51 days left. I'm getting too short to be out there doing this kind of stuff. If I'd been one of those grunts, I never would have climbed into one of those airplanes tonight. If they only knew what was going on up in the cockpit. Out there tonight with a dry pair of gloves and within five minutes they were wringing wet with sweat, and so was everyone else. You come back totally exhausted. I was only out there about two hours and it just killed me. I've got a fierce, splitting, tremendous Excedrin headache right now. It's just unnerving, not to mention scary. Well, I guess I've reiterated that point enough.

Oh, one thing. I think you'll recall I told you I sent in an administrative action form to the headquarters of the Marine Corps requesting permission to travel outside the continental U.S., and also for an additional 10 days leave in connection with my PCS orders. Well, I got a message back. It was disapproved. So we drafted up a message to send back, and it was a real lulu. So the skipper thought maybe we should just sit on it a while.

I had written a personal letter to this guy in Marine Corps headquarters. It seems that Lieutenant Retention has gotten to be so much of a problem that they've set up a special board to deal with the personal problems of junior officers. So if you have a problem with the Corps, you write to them and they try to help you out. Well, I wrote this guy and told him I'd met this girl over here that things had gotten pretty serious, that we'd contemplated marriage that she had gone on home, and how she and I had decided it would be best for us to meet under different circumstances before we made any firm plans. I really laid it on thick. I also told them how I had met a German family that summer I worked in Munich and become close to them, how they came to the States to visit, and all that. And how they were constantly inviting me to come visit them. You know, really thick, the whole nine yards.

The final message that the S-1 drafted up for me said I was engaged to the girl and had to go over there to finalize wedding plans. Pretty strong stuff. So we will have to wait to see how much reaction that letter gets. Also, I may head for Marine Corps headquarters when I get home. The skipper knows some people up there, as does Major B.

My old skipper is in Washington now. He's been assigned to headquarters, but he had two heart attacks, so he is now on convalescent leave. But he might be able to go up there to headquarters and put in a word for me. Good old Colonel Smilanich isn't afraid to tackle anybody. He's a good man to have on your side. Since I flew for him and did a good job for him, I'm sure he'd help me out all he could. In fact, Lt. Colonel S. was around when I first me Sabine.

Well, that's about all for now. All I can say is, boy, does the Marine Corps ever blow.

May 15, 1969

Young Tom and I were sitting here writing letters to our respective whore dogs and shooting the shit. Well, Tom is actually writing to his wife. We got to talking and he came out with this term – zoysia grass. I didn't think anyone else in the world knew anything about zoysia grass except Pop. Well, it turns out that Tom once worked for a nursery and he knows some shit about zoysia grass. So I'm going to turn the mic over to Tom, and he has some helpful hints for Pop on how to handle his zoysia grass.

"Hi, Mr. and Mrs. Ace. Ace was telling me you have planted plugs of zoysia grass in your yard. It struck me as amusing because not that many people have zoysia grass. But like Ace said, I worked for a nursery for three summers. We worked out of Bethesda, Maryland, where the rich kids live. We worked for members of Congress and whatnot. They all have zoysia grass which we started planting for them. There are a couple of things about zoysia grass you need to keep in mind. One thing is that it grows so thick, especially in the late summer, that it will kill itself out. You have to be particularly cognizant of the fact that it has to be cut to a certain level. If you cut it too short, it will die because you damage the roots. It doesn't all grow below the earth, a certain amount of root is exposed.

But if you don't cut it short enough, the grass gets too thick and it's impossible to get enough water to it even though you may water it a couple of times a day. So the length of the grass is really important. I would say if you let it grow somewhere a little over an inch, if you keep it about that height, it will do all right.

Another thing we found through experimentation is that the grass will grow much better if you burn it off in the fall. That sounds a little ridiculous, but the grass starts to turn brown in September, and if you burn it, you not only kill off weeds and things, but it thins the grass out and helps it grow better the following spring. All you have to do is just when it starts turning brown and starts drying, drop a match on it and burn it up. It will turn black and brown and stay that way all winter. But you'll come up with more grass and a better strain of grass in the spring. It's like a pruning effect.

If you're hesitant to try that after all the money and effort you've put into it, you might try just burning a small section to see how it works. Of course, if it doesn't grow back, don't come looking for me. But that's what we used to do and it worked really well. As a matter of fact, I've been trying to get my father to plant some on his head...."

The man you're talking to could use some of that.

"Oh really, well. But why don't you try that. As I say, we've had a lot of success with zoysia grass. It's definitely one of your better grasses."

May 17, 1969

Well, hello folks. I'm lying in my rack listening to my stereo through my earphones. According to my calendar, I only have 46 days and one wake-up left. I also figure I've got about 25 more flying days left because the new Operations Officer said something about getting out of the cockpit about 20 days prior to your departure date, which is okay with me.

I haven't done a whole lot of flying this month; I've become pretty adept at skating. I'm flying tomorrow which is the first time I've been inside a cockpit since the 11th, and only the fourth time I've flown this month.

Today, we had a General come aboard the ship and they're trying to get the whole Squadron to the Philippines next time the ship goes back. They were briefing the General on all the reasons why our Squadron should go to the Philippines, or actually to Okinawa and the Philippines. I sure hope the General buys it. It would be great to take the whole Squadron there. We'd be leaving around the middle of June, and not come back until early July. By that time I'd be long gone.

Also, we heard rumors that there's another Squadron coming up within the next week or so, but they're not firm on that, so we really don't know. I believe this battalion we have on board now is going to be sent back to the bush, and we'll be getting a new battalion on board, which is part of the reason for going back to Okinawa. They want to train the new battalion, and they like to use the Squadron to work with them, using the helicopters and so forth.

I got a letter from Germany the other day which was pretty encouraging. I'm still doing everything I can to get free to go over there. It looks like I will be able to get a passport here, which would save me a lot of trouble when I get back home.

I also got a letter from my brother Dave who is still trying to get me an early out so I can attend his graduation. I don't know if he can work it because there's a great shortage of 46 HACs. They will be most reluctant to let me leave the country before my time is up. But it won't hurt to try.

Tonight we were having a discussion about the relative merits of the guys who've been with the Squadron longer compared to the mid-range guys who've joined the Squadron later. Some of the new guys got a lot of hours in the States. One of them just went over 500 hours and very much wants to be a HAC. The only problem is he doesn't have the experience to back up the hours. Most of the flying he has was straight and level. He's never seen a shot fired in anger. He's been in the Squadron maybe two months now, but he hasn't seen anything. We haven't been flying anything but admin runs the last two months, and this guy is pushing to make HAC. I would be very disappointed to see them make him HAC right off. I think a guy should have at least 200 hours in-country before making HAC.

I guess I had about 300 hours in-country before I made it, but the difference was those hours were hard time, combat flying. Very seldom a day went by without us getting shot at. But these new guys haven't been doing that kind of flying. They haven't had to shoot spiraling approaches into hot zones. They haven't had to take externals and spiral into zones. I think they should be given more time in-country before they push them into these responsibilities.

This one guy is very arrogant and thinks he has all the answers. He's broken several airplanes because he didn't do things properly, even though he knows all the answers. And yet he's got 500 hours now, and is pushing to make HAC. He is one guy I'm just waiting for. If I get an opportunity I'm going to step on him at the slightest provocation. I don't mean step on him seriously, but I'll bring the system down on him, put him in a bind. I have a deep dislike for the guy because you can't talk to him, can't reason with him. We try to tell him things, but he just says, ah yeah, you guys have all the experience, yeah, yeah, I don't know anything. But the truth is he doesn't know anything, and how can you tell a guy anything when he won't listen.

The tragic thing is that when this guy does make HAC, he's going to go out and bust his ass, which in itself will be no great loss, but he's going to take his co-pilot and crew with him, and possibly more people if he has any passengers on board. I think for that reason alone that guy should be made to fly with a co-pilot for a while.

One of the guys observed the other day that he could tell the difference between the Veterans and the new guys by looking in their eyes. When he went to Hawaii not long ago to meet his wife, he had been over here about four or five months at the time, she mentioned that she noticed a change in his physical appearance already. I'd be interested to see what you all think when I get home. I feel like I've aged some. You can't weather all this and not have some effect on you. There have been some times I came back feeling like I had aged 20 years at least. I'm sure my personality has changed some, too.

I'm really looking forward to getting home. It's so close now I can almost taste it. I'm really excited. Well, I'm going to close off for now and get some sleep. It's 11:25 now and I've got to fly the first launch tomorrow. So I'll talk to you later.

May 20, 1969

Well, folks, I got back into the combat flying business today. Went out on re-supply on the wing of a new section leader. In fact, it was this guy's first hop as section leader. We're starting to get new people ready to take over when those of us who have been here a while start to go home.

Oh, by the way, I made Division Leader which means I can take out four airplanes now.

So we're getting these guys all ready with experience. The only combat mission we've been flying here lately is the re-supply for the Korean Marines who are south of Da Nang.

I also had a new co-pilot today, a guy who just got here. We took a little bit of fire going into one of the zones, not much, probably just a sniper. But he got off a least four rounds at us. I was flying at the time hauling an external load. I told the guy we were taking fire. The uninitiated can't always tell ground fire from the sound of the blades. We didn't have any guns covering us, so when we went back we took a different route. But it livened up the day. It was the first time I've been shot at since December, I guess. We still didn't take any hits. But still a little bit of combat flying makes me feel like I'm back in the swing of things.

The shorter I get, the more I insist on keeping the controls, I just can't not take them. Still, I really enjoyed flying with this guy because he had a good attitude. The problem we're having with these new-co-pilots that come from the States is that they have 200 or 300 hours in the 46, which is pretty close to 500 hours in total time. Right away they want to be HAC. What they fail to realize is they don't have the combat experience. Stateside flying is totally different from combat flying.

Their attitude is really aggravating; you can't tell them anything. So the upshot of it all is we just don't let them fly and that's bad because they're not getting the experience they need. But the guy who flew with me today has a really humble attitude and is willing to learn. I'm willing to teach anybody who's willing to learn. I'll show him everything I've learned.

We had a good hop and I let him do some things he's never done before.

The guy didn't scare me once. He did a good job. I got about 6.6 hours today and I was back by 1:30. Not a bad day's work.

Well, I'll talk to you later.

June 3, 1969

Well, folks, it's been a while since I spoke to you last, but I haven't really had a lot to tell. Not too much has been happening around here.

About my party. I would suggest Saturday, the 19th of July. According to the best of my calculations here, I'm supposed to get home on the 12th, which is a Saturday. I intend to spend a couple of days in California seeing some of the old guys that are stationed out there before I come back east. So I probably won't get home until about the 16th.

I'll stay around home there until about the 21st, and then head up to D.C. for a day or so, and then I'm heading to Germany. I figure to spend about two weeks there, assuming I get permission to go. If I don't get permission to go, I'll have to hold it to 12 days because I'll have to sneak over.

I did get a passport; managed to screw the government out of it. I got a five-year passport which normally costs $12 for nothing. I had to give the guy a real bullshit story, but he finally bought it. So regardless of whether the Marine Corps gives me permission, I'm going to Germany.

Anyway, I was thinking of maybe having some kind of lawn party so everybody could come and go as they please, and you could invite everyone. I would be willing to kick in a few bucks to set up the bar. Then again, maybe we could have it out at Davy S.'s place because he has ample room for a lot of people, lots of parking spaces and all. Of course, if it rains we'll have to have it inside.

You don't have to do up any special invitations or anything, and you can invite anyone you want. There's only a few people there I really want to see. But I hope you will invite all the good looking women you know of. I could stand to talk to some women even if I don't know them. Someone can introduce me, or I'll introduce myself. No problem.

I'm really looking forward to driving my car. As I've said before, I sure hope you can have it waiting for me when I land at the airport so I can drive it home.

As far as Dave's comment about C-130s, tell him yes, I did request them, but I got denied like everyone else. The fact that I've served in Vietnam and 10 cents will get me a cup of coffee anywhere in the country, but it won't get me a transition to Fixed Wing. That means that I'm one of that select group of 46 pilots of which there is a great shortage. As a result, I'm going to be flying 46s for a while.

I told you I had written to the Junior Officers Retention Board. I got a letter back explaining why I couldn't have permission to go to Germany. It was not a very good explanation, in my opinion. He also looked in my record and discovered that I had submitted an A.A. for transition to C-130s. He said that in the latter part of 1970, if I still desired transition, I should reapply, and he was sure my application would be seen in a more favorable light. However, by the end of 1970, I'll only have a year left to do, and then I'm getting out. I will not under any circumstances extend, which is what I would have to do. I'd have to guarantee them an additional three years. There's no way I would do that for them because there's no way they'd guarantee I could stay in C-130s. I could be transitioned, sure enough, and just as quickly put back in 46s and sent back over here.

There's no way I'm going to do that. So the Marines can take their 130s and fly them right up their….

"Control yourself."

Tom wants me to control myself. Well, the Marine Corps will get out of me all they can get out of me for the remainder of my tour, and then that's it. I've had enough of these screw jobs that they call bennies.

I talked to York. He's going back to New River. Well, not actually New River. He got a screw job, really. Another outstanding example of the way the Marine Corps watches after their junior officers. York did a tour over here as a FAC, which is a ground tour. It meant that for three months he did no flying. Now, he's got orders back to the Second Marine Division which means he will spend another two years in the States in a ground billet.

So anyway York and I will be in New River together. And more than likely, since he's in a ground billet, he's going to have to come over and get his flight time. So there's a good possibility he and I will be able to fly together in cross countries.

Another guy down here named Steve got orders to new River. Also, last night he got a Dear Steve letter from his girl. I think he and I will be living together with York, the three of us. We should have enough capital to get set up pretty well in terms of living arrangements. We're all pretty much in the same boat.

Tom just said we were all worthless, but Tom is a douchebag. It looks now as though I'm going to be going to Okinawa with the ship. If that's the case, we leave here on the 17th of June, I come back to Vietnam on July 6, and catch a flight out of Da Nang on July 7. I should get back to California by the 12th. And as I said earlier, I should be home by the 16th or maybe the 15th.

Well, all for now.

June 7, 1969

Hi folks. This is Saturday and it hasn't been a very good day for me. I want to finish this tape up and get it off to you, and I just have a little left.

I don't know if I told you, but I have recently been assigned as Basic Training Officer for the Squadron. I'm responsible for the basic training of the enlisted people. They're supposed to maintain a certain level of training throughout their Marine Corps careers, however short those careers may be. In addition, I am also required to work with technical training.

Well today I was supposed to have had the first of two lectures and a movie. One was to be the ship's chaplain, who was going to give a moral guidance lecture. The other was supposed to be a lecture on narcotics to be given by our Squadron flight surgeon. He was going to use a movie too.

We were to use the hangar base as a classroom, and it was just a complete madhouse. There were two rifle platoons having Seabag inspections. There were a couple of basketball games going on. There were forklifts driving around. It was just impossible to get anything done. We stuck it out, but we had poor attendance. When I tried to talk to some of the other people, I got a ration of shit from the Maintenance Officer who is a Major, informing me we didn't have enough airplanes for the training.

The thing that pisses me off is that I'd been given the responsibility by the skipper to conduct this training, but I couldn't get cooperation from anybody. I've about decided to hell with it. If they don't care, I don't care either. My care quotient is at an all-time low.

Epilogue

Those of us who fought in the Vietnam jungles — who bled, suffered and watched our friends die — share an experience that is difficult to decipher, and whose implications are still hotly debated. It was, I am convinced, the definitive experience that has shaped the mindset of my generation. Many of us have struggled for a long time to come to terms with the legacy of Vietnam, and some of us have dealt with it better than others. In my case, it took a long time.

Committing this story to paper has helped me come to terms with the war, and to put it in focus. It has become commonplace to describe the conflict there as a defeat, but those of us who took part in it have difficulty with that. We were never bested on the battlefield. Every time the enemy stood up to challenge us, we took the worst he had to offer and returned it with interest.

It is generally held that the celebrated Tet Offensive was a great communist victory that proved to be the turning point in the war. The reality on the field was that Tet was a debacle for the Viet Cong, after which they were never again able to maintain a viable presence in the field. One more such victory, and the North Vietnamese would have likewise been reduced to impotence.

Yet we were the ones who withdrew, not because we were defeated on the battlefield, but because of a failure of will among our political leadership. For a variety of reasons, the Vietnam conflict provoked a divisive cataclysm that ripped our society apart, and shook our nation to its very foundation.

For me, the final evacuation was terrible humiliation. I thought of the brave men who had given their lives in the cause. Turning our backs and running away seemed to dishonor their sacrifice. It ran contrary to everything I had ever held fast and believed in, a betrayal of my values and heritage.

It was a long time between the evacuation of Saigon and the invasion of Iraq. In the autumn of 1990, as the events in the Persian Gulf unfolded, it seemed increasingly obvious we would become involved militarily. I had no appetite for war, but the issues seemed boldly framed in stark relief. It was as just and clear a cause as ever a nation was called upon to shoulder.

Within that context, I viewed Operation Desert Shield — later Desert Storm — as a grim but badly needed opportunity to confront the scars Vietnam had left on our national psyche. If we were ever going to deal with it, and put that earlier trauma behind us, this was the time.

At that juncture, I was no longer a strapping young Marine pilot, but a mature Reserve Marine Colonel. Though I was a member of the State legislature of Maryland, and though the legislature was then in session, I felt I had to go. So I volunteered for active duty, and was called up.

Of course, it is fitting that men like myself, who are trained and experienced in war, should accept that duty when it arises. Yet my motivation was deeper than that. It seemed to me a last opportunity to relive a transitional period in my life, a time when lines were clearly drawn and every moment seethed with intensity. Other veterans have testified that amid the horrors and deprivation of war, they somehow felt more alive than before or after. In war it is a commonplace that men removed from the most dreadful combat hurry back from their hospital beds or home leave to rejoin their comrades as quickly as they can, drawn like moths to a flame.

My particular flame had to content itself with routine support duties Stateside. The war was over before I could wrangle a combat assignment.

Still, I have gained sufficient insight into life to understand that the real battles are fought, not in trenches or around landing pads, but in legislatures and public debates. Time may have withered my ability to make lightning decisions under heavy enemy fire, but it has at the same time enhanced my capacity to recognize grand folly promoted in the guise of social progress.

The great danger facing our country today is not Asian guerillas, or nuclear missiles, but interior rot caused by civic indifference and political malpractice. As Pogo said, "We have met the enemy and he is us." Our most formidable challenge, and one that comes soon or late to every great nation, is to reaffirm those values of sacrifice, duty, hard work and responsibility that our greatness was born of. To that cause, whose outcome is by no means certain, I have committed my career in public service.

Chapter Photo Descriptions

1 – Ace tape recording from his hootch in Vietnam.

2 – Outdoor disposal point.

3 – Fortifying bunker underneath.

4 – Waiting to go fly.

5 – Ace on the way to insert some recon teams.

6 – Wounded by enemy fire during a routine re-supply mission.

7 – In the cockpit.

8 – Ace as the Coffee Mess Officer playing cards for sodas.

9 – At the office.

10 – First flight after making HAC.

11 – Returning back to Vietnam after Crew Rest in Hong Kong.

12 – Ace in his jungle hat, bullet bouncer vest and carrying his helmet.

13 – Pointing to bullet holes from taking on fire during a flight mission.

14 – Doing a pre-flight check on the Buckeye Bomber.

15 – Aboard the USS Tripoli.

16 – Standing with Captain Shinn, a Korean Marine.

17 – Grooming a handlebar moustache.

18 – Making history; Ace made the 18,000th landing on the USS Tripoli.

19 – Pointing to bullet holes from taking fire during another mission.

20 – Looking out of his helicopter.

21 – Ace just dog tired and ready to finish up his tour and head back to the States.

Biography

Colonel John Chandlee Astle, U.S.M.C. (Ret.)
State Senator, Annapolis, Maryland (Ret.)

Love of country has guided John Astle throughout a career that has included service in the military, law enforcement, and the Maryland General Assembly. John was born in Charles Town, West Virginia on March 31, 1943. A year after graduating from Barboursville High School, Barboursville, West Virginia; John joined the United States Marine Corps Reserve in 1961. He was accepted into the Platoon Leaders Class Program and went on to earn a Bachelor of Arts from Marshall University. Upon graduation in 1966, John joined the United States Marine Corps and received his commission as a Naval Aviator Officer flying helicopters.

John served in the U.S. Marine Corps, Captain, 1966-75. It included serving in Vietnam as a helicopter pilot, 1968-69 and a tour as a Marine One Presidential helicopter pilot, 1971-74. His military service commendations include: the Legion of Merit, two Purple Hearts, the Meritorious Service Medal, 31 air medals and the Presidential Service Badge.

After resigning his regular commission in 1975, John began a 21-year commitment with the Marine Corps Reserve. During that time, he volunteered and served five months active duty in the Persian Gulf War, Desert Storm. He retired as a Colonel from the U.S. Marine Corps Reserve in 1996.

John flew helicopter patrols for the Baltimore Police Department December 1978 through December 1984 then worked as a medical evacuation pilot for the Washington Hospital Center from 1985-2005.

As an American politician, John continued his public service by serving the people of Maryland first as a member of the Maryland House of Delegates representing District 30, Anne Arundel County, January 1983 to January 11,

1995, and then as a member of the Maryland Senate, January 11, 1995 to January 9, 2019. A respected leader, John served as Vice-Chairman of the powerful Senate Finance Committee, where he played a key role in shaping state policy on energy, healthcare, banking and insurance. He also served as Chairman of the Anne Arundel County Senate Delegation, and as a member of the Joint Audit Committee, the Legislative Policy Committee and the Joint Information Technology and Biotechnology Committee. An avid sportsman and environmental advocate, John is a long-time supporter of the National Congressional Sportsmen's Foundation. He spearheaded the creation of the Maryland Legislative Sportsmen's Caucus in 2001, which led to the creation of a National Assembly of Sportsmen's Caucuses. He was a Senate Chair of the Maryland Caucus and a Past President of the national organization.

John also served in key leadership positions with organizations working to restore the Chesapeake Bay, including a position on the Chesapeake Bay Trust's Board of Trustees. John is a member of the American Legion, Veterans of Foreign Wars, Disabled American Veterans, National Rifle Association, Safari Club International and was an active member of the Maryland Veterans Caucus. He served as a Member of the Board of the Maryland Tourism Development Board, and Chairman of the Board of Visit Annapolis & Anne Arundel County. John lives in Annapolis, Maryland with his wife Jayne.

Author Bio

Gemma is the daughter of creative and entrepreneurial first generation Italian/American parents. Growing up as the youngest of 4 and spending her youth working in the family restaurant business, she learned a lot about business and people. After graduating college, she enjoyed a career in commodity futures trading for nineteen years while living in Manhattan and Chicago. On a snowy evening visiting her sister, she met the man of her dreams. Soon after, Gemma relocated and married Mike off the 18th hole on Pebble Beach in Monterey, California. She happily and creatively lives in Southeastern Pennsylvania and writes books about people and circumstances. After meeting and becoming friends with John over the past 18 years, Gemma was and remains impressed with his military stories about his heroism and that of his military brothers. She felt compelled to put his words to print, and to share with others who wore the uniform, who served, and who made the ultimate sacrifice defending our freedom.